# Progressive Agriculture

CONTENT VIBES
PUBLISHERS • PRINTERS • AGGREGATORS

# About the Authors

**Dr. Arabinda Dhal** passed M.Sc. (Ag) in 1991 from OUAT and recipient of Ph.D. degree from UTKAL University, Bhubaneswar during 2015. Mycology and plant pathology is his area of specialization. He has been working in OUAT both in teaching, extension and research for more than 20 years. He is a life member of Indian Mycological Society, Association of Rice Research Workers, NRRI & Society of plant protection and environment, Bhubaneswar. He attended many seminar, workshops; presented papers & awarded. He developed 5 Technologies in forage pathology n received ICAR appreciation award ... PUBLICATION ... Dr Dhal has more than 30 research publications in National and International journals of repute to his credit by date.

**Dr. Kavya Thottempudi** passed out B.Sc. Agriculture from Annamalai University, Chidambaram during 2016. She pursued Masters by working on Stability of selected Blackgram (*Vigna mungo* (L.) Hepper) genotypes across environments represented by dates of sowing at University of Agricultural Sciences, College of Agriculture, GKVK, Bangalore, Karnataka in Agriculture specialization in Genetics and Plant Breeding during 2018 under the guidance of Dr. S. Rangaiah. After her post-graduation she worked as Teaching Associate in Darsi Agriculture college, under ANGRAU, LAM, Guntur, Andhra Pradesh. During time as Teaching associate, I worked as an advisor for Participatory Agricultural Management Programme (PAMP) and also worked as university examiner and attended examination duties. Later she pursued Ph.D. degree by working on Comparative assessment of commercial grain yield and seed production potentials of single cross hybrids (SCH) and modified single cross hybrids (MSCH) in maize (*Zea mays* L.). at University of Agricultural Sciences, College of Agriculture, GKVK, Bangalore, Karnataka during the year 2019 under the guidance of Dr. J. Shanthala. Memberships and Awards:She is a life member in Indian Science Congress, Indian Society of

Agricultural and Horticultural Research and Development. She received many awards in seminars and three awards in conferences and also best Ph.D. thesis award during doctoral studies. Publications: Credited with publication of three research papers, one review paper, four abstracts and three popular articles.

**Dr. Sanjay Koushal**, presently working as Scientist, Agronomy at Krishi Vigyan Kendra, Reai of SKUAST, Jammu, India. He was born on 15 Sept., 1977 in a Kathua district of Jammu & Kashmir. He is a very young, energetic and dedicated scientist. He did his B.Sc. Ag and M.Sc. (Agronomy) Khalsa College, GNDU, Amritsar, Punjab and did his Ph.D. (Agronomy) from Sher-e-Kashmir University of Agricultural Sciences & Technology of Jammu (J&K), India in 2008. He has joined as Scientist Agronomy in KVK, Reasi, SKUAST-Kashmir in June, 2014. Before Joining as scientist in KVK, Reasi he was Research Associate in Agromet Research Centre SKUAST Jammu and worked on Crop Modeling Climate Change, Resilient Agriculture and Agro Adviosry services at district levels. Dr Sanjay Koushal has vast experience and expertise in Crop Modeling, Climate Change and their effect on Field crops and soil fertility He has also developed a Low Cost Vermicompost model for small and marginal farmers of mid hills of Jammu, Popularised Specific Integrated Farming System Model for Mid Hills of Reasi and Udhampur district. He was awarded with Life Fellow of “The Agrometerological Accociation of India, Gujarat” on 2011. He is an Associate Editor and reviewer of “International Journal of Agriculture Sciences” and "Environmental Science Archives”. He has handled several externally funded as well as internally funded projects as a P.I and Co-P.I. He has several publications to his credit which includes research papers, technical bulletins, extension folders, practical manuals and book chapters published in reputed national/international journals and books. He had conducted several on-spot interactive Programmes, Farmers Skill Development Trainings, workshops, awareness programmes and Model Training Cources on Climate Change and variability, IFS, Vermicompost Producers. He was also incharge KVK Farm with an area of 270 kanals and Vermicompost unit at KVK with annual production of 10 tonnes per annum. He was awarded with Best Paper/Posters presentations in national conferences and act as Rapporteur in conferences. He has evaluated and set numerous papers as an external examiner. His research work has been widely appreciated and referred in national and international research papers. He has delivered several expert lectures on Soil fertility , Integrated nutrient Management of Field crops, Vegetable crops, oilseed crops, cereal crops and Integrated Farming System, Conservation Agriculture and Natural Farming to Junior Agriculture Officers,

Field functionaries, Extension workers and farmers to promote and enhance the adoptability of Integrated Pest Management. He has delivered many TV and radio talks (AIR). He is a certified trainer of Agriculture Skill Council of India (ASCI). He has conducted several farmers training programmes including 5 days vocational training programmes on Entrepreneurship in Agricutlure and allied sectors also Coordinated (200 hrs)/ 35 Days Skill Development Training on Vermicompost producer, Nursery Raising under ASCI and several 15 days training for Fertilizer dealers under INM trainings. He also coordinator of 17 days residential training Programme of Pashu Sakhis A-Help Training sponsored by NDDB, Gujarat at Directorate of Extension SKUAST Jammu and Nodal Officer of Agri Clinic and Agri Business (ACABC) By MANAGE Hyderabad.

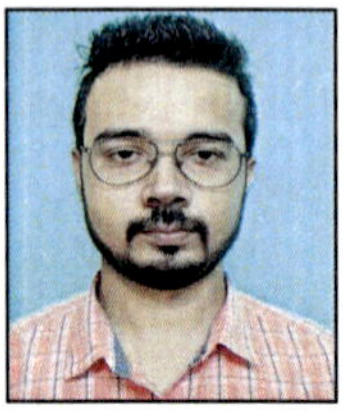

**Priyam Goswami** did his B.Tech in Agricultural Engineering from School of Engineering and Technology, Nagaland University in 2021. He completed M.Tech in Farm Machinery and Power at North Eastern Regional Institute of Science and Technology, Arunachal Pradesh in the year 2023. He was awarded with the gold medal for academic excellence in the masters degree. He has the experience of working at the Department of Agricultural Engineering of Assam Agricultural University, Jorhat which is one of the most prestigious agricultural universities of India. Currently he is pursuing Ph.D. in North Eastern Regional Institute of Science and Technology, Arunachal Pradesh. He has attended many national level workshops and seminars in various reputed institutes and universities of India. He is currently doing research on precision agriculture technologies and has expertise in robotics and its application in the agricultural sector.

**Dr. P. Kalaiselvi**, Associate Professor (Env. Science) handled under graduate courses, post graduate courses and doctoral course. She has guided 15 postgraduate students. As a Researcher, she has worked in two core themes of Environmental Science Department including Bioremediation and Solid waste management. Developed a sustainable technology for bioremediation of industrially polluted ecosystems and also developed microbial consortium for lignocellulosic waste management. Presently she is working in the in situ decomposition of rice stubbles and developing an Effective Microbial formulation product for managing both liquid and solid waste. She served as principal investigator in venture capital scheme operating in the Department of Env. Science which focuses on production of bioinputs like TNAU Biomineralizer, TNAU Panchagavya & Biocompost and supplying for various stakeholders. She also looked after the compost technology Park in Tamil Nadu Agricultural University where the organic manure production from biodegradable waste was done and also monitored various waste management activities. In the part of extension and exhibition incharge she organized various exhibitions to portray the technologies developed by the Department for farmers welfare. She alos involved in the media popularization of the technologies, attended various trainings, seminars, symposium and workshops to improve my professional skills and I have also presented papers, mainly to share my knowledge with the peers.Published six books, several chapters in books, research articles with high NAAS rating in various journals of national and international repute.

**Mr. Ritik Raj** is currently serving as a Research Scholar in the Department of Botany, Plant Physiology, and Biochemistry at the College of Basic Science and Humanities, Dr. Rajendra Prasad Central Agricultural University, Pusa, Samastipur. He holds a degree from Tamil Nadu Agricultural University. Mr. Raj has made significant contributions to the field through his authorship of numerous articles, books, and book chapters published in both National and International Journals. He has also participated in various training programs within the biotech industry and other esteemed institutions. His research interests are focused on Plant Abiotic Stress Tolerance in millet and the screening of genotypes for waterlogging tolerance in millet.

# Progressive Agriculture

Arabinda Dhal
Kavya Thottempudi
Sanjay Koushal
Priyam Goswami
P. Kalaiselvi
Ritik Raj

CONTENT VIBES
PUBLISHERS • PRINTERS • AGGREGATORS

CV CONTENT VIBES
PUBLISHERS • PRINTERS • AGGREGATORS

1168, Sector 13, Urban Estate
Karnal-132 001, Haryana
Tel: 91-84470 75807, 18440 41168
Email: contentvibesppa@gmail.com
www.contentvibes.in



Print ISBN: 978-81-97781-57-5
e-book ISBN: 978-81-97781-58-2

# Perface

Agriculture, the cornerstone of human civilization, has continually evolved to meet the growing demands of a changing world. In an era where the challenges of climate change, population growth, and resource scarcity are ever-present, the need for progressive approaches to agriculture has never been more critical. This book, "Progressive Agriculture," explores the innovative strategies and cutting-edge technologies shaping the future of farming.

At its core, progressive agriculture represents a shift from traditional practices towards more sustainable and efficient methods. It encompasses a wide range of advancements, from precision farming and agroecology to genetic improvements and advanced irrigation systems. The aim is not only to increase productivity but also to enhance environmental stewardship, ensure food security, and promote the well-being of farming communities.

In the chapters that follow, we will delve into the principles and practices that define progressive agriculture. We will examine how modern technologies are being harnessed to address the challenges faced by farmers worldwide, and we will highlight the stories of those who are leading the way in this transformative field.

This book is intended for farmers, researchers, policymakers, and anyone with a stake in the future of agriculture. By understanding and embracing the principles of progressive agriculture, we can collectively work towards a more resilient and sustainable food system that will benefit generations to come.

**Authors**

# Contents

# 1

# Introduction to Progressive Agriculture

**Abstract**

*Progressive agriculture represents the intersection of innovative practices and sustainable farming techniques aimed at improving productivity, environmental stewardship, and social responsibility in agriculture. As global challenges such as climate change, population growth, and resource scarcity intensify, the need for a progressive approach to agriculture becomes ever more critical. This chapter delves into the principles, practices, and benefits of progressive agriculture, tracing its evolution from traditional farming methods to modern, technology-driven approaches. The discussion covers key areas such as sustainable farming practices, precision agriculture, biotechnology, and the role of policy and education in advancing progressive agriculture. By integrating scientific advancements with ecological principles, progressive agriculture seeks to create resilient agricultural systems capable of sustaining future generations.*

***Keywords***: *Progressive agriculture, sustainable farming, precision agriculture, biotechnology, climate change, food security, agricultural innovation, environmental stewardship, policy in agriculture, agricultural education.*

## Introduction

Agriculture, the backbone of human civilization, has undergone significant transformations over millennia. From the domestication of plants and animals to the industrialization of farming, each stage of agricultural evolution has been marked by advances that have shaped societies. However, the 21st century presents unique challenges that require a reevaluation of traditional agricultural practices. Climate change, population growth, and dwindling natural resources necessitate a shift towards more sustainable and innovative farming methods—what is now termed "progressive agriculture."

Progressive agriculture is not merely a set of practices but a holistic approach that integrates modern technology, sustainable practices, and socio-economic considerations to create an agricultural system that meets the demands of today while ensuring the well-being of future generations. This chapter explores the concept of progressive agriculture, its historical context, key components, and its potential to address global food security and environmental challenges.

## Historical Context of Agriculture

The history of agriculture is one of continuous adaptation and innovation. The Neolithic Revolution, which began around 10,000 BCE, marked the transition from hunter-gatherer societies to settled agricultural communities. The development of irrigation systems, crop rotation, and animal husbandry laid the foundation for modern agriculture.

The 18th and 19th centuries saw the advent of the Agricultural Revolution, characterized by the introduction of machinery, improved crop varieties, and the use of chemical fertilizers. This period significantly increased agricultural productivity, laying the groundwork for the Green Revolution of the mid-20th century. The Green Revolution, driven by advances in plant breeding, chemical fertilizers, and irrigation, led to unprecedented increases in crop yields, particularly in developing countries.

However, the Green Revolution also brought about significant environmental and social challenges, including soil degradation, water pollution, and the marginalization of small-scale farmers. These issues highlighted the need for a more sustainable approach to agriculture, paving the way for the concept of progressive agriculture.

## Defining Progressive Agriculture

Progressive agriculture is an approach that combines scientific innovation with sustainable farming practices to create an agricultural system that is productive, environmentally sound, and socially equitable. It is characterized by several key principles:

1. **Sustainability:** Ensuring that agricultural practices do not deplete natural resources or harm the environment, thereby maintaining the ability of future generations to meet their needs.
2. **Innovation:** Leveraging advances in technology, such as precision agriculture and biotechnology, to enhance productivity and reduce the environmental impact of farming.
3. **Resilience:** Building agricultural systems that can withstand and adapt to challenges such as climate change, pests, and diseases.
4. **Equity:** Promoting social and economic justice within the agricultural sector, ensuring that all stakeholders, including smallholder farmers and rural communities, benefit from agricultural advancements.
5. **Education and Awareness:** Fostering a culture of continuous learning and knowledge sharing among farmers, policymakers, and the public to support the adoption of progressive practices.

## Key Components of Progressive Agriculture

### 1. Sustainable Farming Practices

Sustainable farming practices form the bedrock of progressive agriculture, aimed at creating systems that are both productive and environmentally sound. These practices are designed to conserve resources, enhance soil health, protect biodiversity, and reduce the environmental impact of farming activities. By balancing ecological and economic needs, sustainable farming practices help ensure the long-term viability of agricultural systems. Below are detailed explanations of key sustainable farming practices.

### 1. Organic Farming

Organic farming is a system that seeks to produce food in a natural way, avoiding the use of synthetic chemicals such as pesticides, herbicides, and synthetic fertilizers. Instead, organic farmers rely on biological processes, natural inputs, and organic fertilizers to maintain soil fertility and manage pests and diseases. The principles of organic farming are based on the concept of working with nature rather than against it.

- **Soil Health:** Organic farming places a strong emphasis on soil health, recognizing that healthy soils are the foundation of productive agriculture. Techniques such as composting, green manuring, and the use of organic amendments like animal manure and cover crops help improve soil structure, increase organic matter content, and enhance nutrient availability.
- **Biodiversity:** Organic farms often feature a high level of biodiversity, both in terms of the variety of crops grown and the presence of beneficial organisms. This biodiversity helps create a more resilient farming system, reducing the reliance on chemical inputs to control pests and diseases.
- **Pest and Disease Management:** Instead of using synthetic pesticides, organic farmers employ a range of strategies to manage pests and diseases. These include crop rotation, intercropping, the use of biological controls (such as predators and parasitoids), and the application of natural pesticides like neem oil or insecticidal soaps.
- **Certification and Standards:** Organic farming is governed by strict standards and certification processes, which vary by country but generally prohibit the use of genetically modified organisms (GMOs) and synthetic chemicals. Certification provides assurance to consumers that products labeled as organic meet these standards.

## 2. Agroecology

Agroecology is an approach to farming that applies ecological principles to the design and management of agricultural systems. It emphasizes the integration of crops, livestock, and natural habitats to create a diverse and sustainable farm ecosystem. Agroecology is not only about production but also about creating a harmonious relationship between agriculture and the environment.

- **Ecological Balance:** Agroecology aims to enhance the ecological balance on farms by promoting biodiversity, recycling nutrients, and using renewable resources. Techniques such as polycultures (growing multiple crops together), agroforestry (integrating trees with crops and livestock), and natural pest management are central to agroecological farming.
- **Resilience:** By mimicking natural ecosystems, agroecology creates more resilient farming systems that can better withstand environmental stresses such as drought, pests, and diseases. The diversity of species and interactions in agroecological systems helps buffer against these challenges.
- **Social and Cultural Aspects:** Agroecology also considers the social and cultural dimensions of farming, recognizing the importance of traditional knowledge and practices. It promotes local food systems, food sovereignty, and the empowerment of smallholder farmers and rural communities.
- **Economic Viability:** While agroecology focuses on ecological sustainability, it also seeks to ensure the economic viability of farming by reducing dependence on costly external inputs and enhancing farm productivity through ecological means.

## 3. Conservation Agriculture

Conservation agriculture is a set of soil management practices that minimize disruption to the soil's structure, composition, and natural biodiversity. The main principles of conservation agriculture are minimal soil disturbance, maintaining a permanent soil cover, and diversifying crop species. These practices help protect the soil from erosion, improve water retention, and increase carbon sequestration.

- **No-Till Farming:** No-till farming, or zero tillage, is a method where the soil is left undisturbed from harvest to planting, except for the planting holes. This practice reduces soil erosion, enhances soil organic matter, and conserves moisture. By avoiding plowing, no-till farming also helps maintain soil structure and reduces the loss of organic carbon.

- **Cover Cropping:** Cover crops, such as clover or rye, are planted during off-seasons when the soil would otherwise be bare. These crops protect the soil from erosion, suppress weeds, and improve soil fertility by fixing nitrogen or adding organic matter when they decompose.
- **Crop Rotation:** Rotating crops from season to season helps break pest and disease cycles, improves soil health, and reduces the need for chemical inputs. Different crops have different nutrient requirements and contributions, so rotating them helps balance soil nutrients and prevent depletion.
- **Permanent Soil Cover:** Maintaining a permanent cover on the soil, whether through crop residues, cover crops, or mulch, protects the soil from erosion, reduces water evaporation, and enhances soil organic matter. It also provides habitat for beneficial organisms such as earthworms and microorganisms that improve soil fertility.

### 4. Integrated Pest Management (IPM)

Integrated Pest Management (IPM) is an ecological approach to pest control that combines a variety of management strategies and practices to minimize the use of chemical pesticides. The goal of IPM is to control pest populations to a level that does not cause economic harm while minimizing risks to human health and the environment.

- **Monitoring and Identification:** The first step in IPM is regular monitoring and accurate identification of pests. This allows farmers to make informed decisions about whether pest control is needed and what methods to use.
- **Preventive Practices:** IPM emphasizes preventive practices that reduce the likelihood of pest outbreaks. These include crop rotation, use of resistant varieties, habitat management, and maintaining healthy crops through proper nutrition and irrigation.
- **Biological Control:** Biological control involves using natural predators, parasitoids, or pathogens to control pest populations. For example, ladybugs can be introduced to control aphid populations, and parasitic wasps can be used to target caterpillar pests.
- **Mechanical and Physical Controls:** Mechanical controls include traps, barriers, and hand-picking of pests, while physical controls involve altering the environment to make it less favorable for pests, such as using row covers to protect plants from insects or adjusting planting times to avoid peak pest periods.

- **Chemical Control:** When necessary, IPM allows for the use of chemical pesticides, but they are applied in a targeted and judicious manner. This minimizes the impact on non-target organisms and reduces the risk of pesticide resistance. Often, IPM programs recommend using the least toxic options available.

### 5. Agroforestry

Agroforestry is the practice of integrating trees and shrubs into agricultural landscapes. This approach combines agriculture and forestry to create more diverse, productive, and sustainable land-use systems. Agroforestry has numerous benefits, including enhancing biodiversity, improving soil health, sequestering carbon, and providing additional sources of income for farmers.

- **Silvopasture:** This is the integration of trees and shrubs with livestock grazing. Trees provide shade and shelter for animals, improve soil health through leaf litter and root systems, and can produce additional income through timber, fruit, or nuts.
- **Alley Cropping:** Alley cropping involves planting rows of trees or shrubs between rows of crops. The trees can provide windbreaks, reduce soil erosion, and improve nutrient cycling. In some systems, nitrogen-fixing trees are used to enhance soil fertility for the crops grown in the alleys.
- **Forest Farming:** Forest farming is the cultivation of high-value specialty crops such as mushrooms, medicinal plants, or nuts under the canopy of an existing forest. This practice utilizes the shade and microclimate provided by the forest while maintaining forest cover and biodiversity.
- **Windbreaks:** Planting rows of trees or shrubs as windbreaks can protect crops from wind damage, reduce soil erosion, and provide habitat for wildlife. Windbreaks can also contribute to carbon sequestration and improve the microclimate for adjacent crops.

### 6. Water Conservation and Management

Efficient water use is crucial for sustainable farming, particularly in regions where water is scarce. Water conservation practices aim to reduce water usage, improve irrigation efficiency, and protect water resources from pollution and overuse.

- **Efficient Irrigation Systems:** Drip irrigation and sprinkler systems are more efficient than traditional flood irrigation, as they deliver water directly to the plant roots, reducing evaporation and runoff. These systems can be further optimized with soil moisture sensors and automated controls.

- **Rainwater Harvesting:** Collecting and storing rainwater for agricultural use can reduce reliance on external water sources and improve water availability during dry periods. Rainwater harvesting systems can range from simple barrels to more complex catchment and storage systems.
- **Soil Moisture Management:** Practices such as mulching, cover cropping, and no-till farming help retain soil moisture by reducing evaporation and improving soil structure. These practices also enhance the soil's ability to absorb and store water.
- **Water Recycling and Reuse:** Reusing water from agricultural runoff, treated wastewater, or other sources can reduce the demand for fresh water. In some cases, constructed wetlands are used to treat wastewater before it is reused for irrigation.

### Benefits of Sustainable Farming Practices

The adoption of sustainable farming practices offers numerous benefits, including:

- **Environmental Protection:** Sustainable farming practices reduce soil erosion, conserve water, enhance biodiversity, and minimize pollution. By protecting natural resources, these practices help maintain ecosystem services that are vital for agriculture and human well-being.
- **Climate Change Mitigation:** Practices such as no-till farming, agroforestry, and cover cropping sequester carbon in soils and vegetation, helping to mitigate climate change. Sustainable farming also reduces greenhouse gas emissions by minimizing the use of synthetic inputs and improving energy efficiency.
- **Economic Resilience:** Sustainable farming practices can improve farm profitability by reducing input costs, increasing productivity, and diversifying income sources. By enhancing the resilience of farming systems to environmental and economic shocks, these practices contribute to long-term economic sustainability.
- **Social and Community Benefits:** Sustainable farming practices often promote social equity by supporting smallholder farmers, preserving traditional knowledge, and fostering local food systems. They also contribute to food security by increasing the availability of healthy, nutritious food produced in environmentally friendly ways

## 2. Precision Agriculture

Precision agriculture is a technology-driven approach that uses data and advanced tools to optimize agricultural practices. By applying precise amounts of inputs such as water, fertilizers, and pesticides, farmers can reduce waste, lower costs, and minimize environmental impact. Key technologies in precision agriculture include:

- **GPS and GIS:** Global Positioning Systems (GPS) and Geographic Information Systems (GIS) allow farmers to map their fields, monitor crop health, and plan operations with precision.
- **Remote Sensing:** Drones, satellites, and sensors collect data on crop conditions, soil moisture, and pest activity, enabling farmers to make informed decisions.
- **Variable Rate Technology (VRT):** Equipment that applies inputs at variable rates across a field based on the specific needs of different areas, improving efficiency and reducing input use.
- **Automated Machinery:** Tractors, harvesters, and other machinery equipped with automation and AI can perform tasks more accurately and efficiently than human-operated equipment.

## 3. Biotechnology in Agriculture

Biotechnology is another critical component of progressive agriculture, offering tools to enhance crop yields, improve nutritional content, and increase resistance to pests, diseases, and environmental stresses. Key biotechnological advancements include:

- **Genetically Modified Organisms (GMOs):** Crops that have been genetically engineered to possess desirable traits, such as herbicide tolerance, pest resistance, or improved nutritional content.
- **CRISPR and Gene Editing:** Advanced techniques for precisely altering the genetic makeup of crops and livestock, potentially speeding up the development of new varieties and breeds with improved traits.
- **Biopesticides and Biofertilizers:** Natural or engineered organisms and compounds that can be used to control pests and enhance soil fertility, reducing the reliance on synthetic chemicals.

## 4. Climate-Smart Agriculture

Climate change poses one of the most significant threats to global agriculture, with rising temperatures, changing precipitation patterns, and more frequent extreme weather events. Climate-smart agriculture (CSA) aims to address these challenges by promoting practices that increase resilience to climate change while reducing greenhouse gas emissions. Key strategies include:

- **Crop Diversification:** Growing a variety of crops to spread risk and increase resilience to climate-related stresses.

- **Water Management:** Implementing efficient irrigation systems, such as drip or sprinkler irrigation, to conserve water and reduce vulnerability to drought.
- **Agroforestry:** Integrating trees and shrubs into agricultural landscapes to enhance biodiversity, sequester carbon, and provide additional sources of income.
- **Carbon Sequestration:** Practices such as cover cropping, reduced tillage, and improved pasture management that increase the amount of carbon stored in soils and vegetation.

### 5. Policy and Education

For progressive agriculture to be widely adopted, supportive policies and educational initiatives are essential. Governments, NGOs, and international organizations play a crucial role in creating an enabling environment for progressive practices. Key areas of focus include:

- **Subsidies and Incentives:** Providing financial support to farmers who adopt sustainable practices, invest in new technologies, or transition to organic farming.
- **Research and Development:** Funding research into new agricultural technologies and practices, as well as supporting the development of locally adapted solutions.
- **Extension Services:** Offering education and training to farmers on progressive practices, including the use of new technologies, sustainable farming methods, and climate adaptation strategies.
- **Land Tenure and Access:** Ensuring that farmers have secure land tenure and access to the resources they need to implement progressive practices.

## The Future of Progressive Agriculture

As the global population continues to grow, and the impacts of climate change become more pronounced, the importance of progressive agriculture will only increase. Future advancements in technology, coupled with a deeper understanding of ecological processes, will likely lead to even more sophisticated and sustainable farming systems.

Some emerging trends that could shape the future of progressive agriculture include:

- **Regenerative Agriculture:** A holistic approach that focuses on restoring and enhancing the health of ecosystems, particularly soil health, through practices such as rotational grazing, agroforestry, and organic farming.

- **Vertical Farming and Urban Agriculture:** Innovative approaches to food production that use limited space and resources, potentially transforming urban landscapes and reducing the carbon footprint of food production.
- **Blockchain and Food Traceability:** Technology that enhances transparency and traceability in food supply chains, helping to ensure the integrity of organic and sustainably produced food.
- **Synthetic Biology:** The design and construction of new biological parts and systems, which could lead to breakthroughs in crop production, pest control, and soil health.

## Challenges and Opportunities

While the potential benefits of progressive agriculture are immense, several challenges must be addressed to realize its full potential:

- **Access to Technology:** Ensuring that smallholder farmers, particularly in developing countries, have access to the tools and knowledge needed to implement progressive practices.
- **Economic Viability:** Balancing the need for sustainable practices with the economic realities of farming, particularly in regions where farmers operate on thin margins.
- **Consumer Awareness:** Educating consumers about the benefits of progressive agriculture and encouraging demand for sustainably produced food.
- **Policy Support:** Developing and implementing policies that support progressive agriculture, including subsidies, research funding, and education programs.

Despite these challenges, the opportunities presented by progressive agriculture are vast. By adopting a forward-thinking approach that integrates technology, sustainability, and equity, the agricultural sector can play a vital role in addressing some of the most pressing challenges of our time.

## Conclusion

Progressive agriculture represents a transformative approach to farming that is essential for meeting the challenges of the 21st century. By embracing innovation, sustainability, and social responsibility, progressive agriculture offers a pathway to a more resilient and equitable food system. As we look to the future, the continued evolution of agricultural practices, supported by sound policy and education, will be critical in ensuring that agriculture can sustain both people and the planet for generations to come.

## References

Bhatt, S., Sharma, S., & Jain, P. (2022). Rural innovation using technology in progressive agriculture: Empowering women towards sustainable rural development. In Driving Factors for Venture Creation and Success in Agricultural Entrepreneurship (pp. 227-240). IGI Global.

Gomase, A. S., & Tekale, V. S. Progressive Research: An International Journal.

Siegel, B. (2017). Modernizing Peasants and "Master Farmers" Progressive Agriculture in Early Independent India. Comparative Studies of South Asia, Africa and the Middle East, 37(1), 64-85.

Sultana, S., Sagar, A., Mia, M. A., Kabir, M. H., Shabi, T. H., Dhar, M., & Hossain, A. K. M. Z. (2021). Progressive Agriculture.

# 2

# Sustainable Farming Practices

**Abstract**

*Sustainable farming practices are at the heart of the global effort to build resilient agricultural systems that can meet the demands of a growing population while preserving the environment. As the world grapples with challenges such as climate change, soil degradation, and biodiversity loss, the adoption of sustainable farming methods becomes increasingly critical. This chapter provides an in-depth exploration of sustainable farming practices, tracing their historical development, key principles, and various approaches. It delves into practices like organic farming, agroecology, conservation agriculture, integrated pest management, and water conservation, illustrating how these methods contribute to environmental health, economic viability, and social equity. The discussion also highlights the benefits, challenges, and future prospects of sustainable farming, emphasizing its role in ensuring food security and sustainability for future generations.*

***Keywords***: *Sustainable farming, organic farming, agroecology, conservation agriculture, integrated pest management, water conservation, soil health, biodiversity, climate change, food security.*

## Introduction

Sustainable farming practices have emerged as a crucial response to the environmental and social challenges facing modern agriculture. These practices aim to create agricultural systems that are environmentally sound, economically viable, and socially responsible. As agriculture continues to face pressures from climate change, population growth, and resource depletion, the need for sustainable approaches has never been more urgent.

## Historical Context of Sustainable Farming

The concept of sustainable farming has deep historical roots, with traditional farming systems around the world often embodying principles of sustainability. Indigenous agricultural practices, such as crop rotation, mixed cropping, and the use of organic fertilizers, were inherently sustainable, relying on ecological knowledge passed down through generations.

The industrialization of agriculture in the 19th and 20th centuries, however, brought significant changes. The Green Revolution, characterized by the widespread adoption of chemical fertilizers, pesticides, and high-yielding crop varieties, dramatically increased agricultural productivity. While this revolution helped alleviate hunger in many parts of the world, it also led to unintended consequences, including soil degradation, water pollution, and the loss of biodiversity.

In response to these challenges, the late 20th century saw the rise of the sustainable agriculture movement, which sought to integrate the benefits of modern agricultural technology with the ecological principles of traditional farming. This movement laid the foundation for the sustainable farming practices we see today.

## Key Principles of Sustainable Farming

Sustainable farming is guided by several core principles that aim to balance the needs of food production with the preservation of natural resources and the well-being of communities. These principles include:

1. **Environmental Stewardship:** Sustainable farming seeks to protect and enhance the natural environment by conserving resources, reducing pollution, and promoting biodiversity.
2. **Economic Viability:** For farming to be sustainable, it must be economically viable, providing farmers with a stable and sufficient income while minimizing the need for costly inputs.
3. **Social Responsibility:** Sustainable farming practices promote social equity, ensuring that all members of society, including smallholder farmers and marginalized communities, benefit from agricultural development.
4. **Resilience:** Sustainable farming aims to build resilient agricultural systems that can withstand and adapt to environmental and economic challenges, such as climate change, market fluctuations, and resource scarcity.
5. **Innovation and Adaptation:** Sustainable farming encourages the adoption of new technologies and practices that improve efficiency and reduce environmental impact, while also respecting and integrating traditional knowledge.

## Key Sustainable Farming Practices

### 1. Organic Farming

Organic farming is one of the most well-known sustainable farming practices. It focuses on producing food without synthetic chemicals, instead relying on natural processes and inputs to maintain soil fertility and control pests. Organic farming is built on the principles of health, ecology, fairness, and care, and is governed by strict standards and certification processes.

- **Soil Management:** Organic farmers prioritize soil health by using practices such as composting, green manuring, and crop rotation. These practices enhance soil fertility, improve structure, and increase organic matter content, which in turn supports healthy plant growth and reduces the need for chemical fertilizers.
- **Biodiversity:** Organic farms typically feature high levels of biodiversity, both in terms of the variety of crops grown and the presence of beneficial organisms. This biodiversity helps create a resilient farming system, reducing the reliance on chemical inputs to control pests and diseases.
- **Pest and Disease Management:** Organic farming employs a range of strategies to manage pests and diseases without synthetic pesticides. These include the use of resistant crop varieties, biological controls (such as predators and parasitoids), and cultural practices like crop rotation and intercropping.
- **Certification:** Organic farming is subject to certification by various organizations, which set standards for the production, processing, and labeling of organic products. Certification provides assurance to consumers that organic products meet these standards and have been produced using environmentally friendly practices.

### 2. Agroecology

Agroecology is an approach to farming that applies ecological principles to the design and management of agricultural systems. It seeks to create diverse, resilient, and sustainable farming systems that mimic natural ecosystems. Agroecology is not only a set of practices but also a social movement that promotes food sovereignty and the empowerment of smallholder farmers.

- **Ecological Balance:** Agroecology emphasizes the importance of ecological balance on farms, promoting biodiversity, nutrient cycling, and the use of renewable resources. Practices such as polycultures, agroforestry, and natural pest management are central to agroecology.

- **Resilience:** By mimicking natural ecosystems, agroecology creates farming systems that are more resilient to environmental stresses, such as drought, pests, and diseases. The diversity of species and interactions in agroecological systems helps buffer against these challenges.
- **Social and Cultural Aspects:** Agroecology also considers the social and cultural dimensions of farming, recognizing the importance of traditional knowledge and practices. It promotes local food systems, food sovereignty, and the empowerment of smallholder farmers and rural communities.
- **Economic Viability:** While agroecology focuses on ecological sustainability, it also seeks to ensure the economic viability of farming by reducing dependence on costly external inputs and enhancing farm productivity through ecological means.

### 3. Conservation Agriculture

Conservation agriculture is a set of soil management practices that aim to minimize soil disturbance, maintain a permanent soil cover, and promote crop diversity. These practices help protect the soil from erosion, improve water retention, and increase carbon sequestration, making them essential for sustainable farming.

- **No-Till Farming:** No-till farming, or zero tillage, involves planting crops without disturbing the soil through plowing or tilling. This practice reduces soil erosion, enhances soil organic matter, and conserves moisture, helping to maintain soil health and productivity.
- **Cover Cropping:** Cover crops, such as clover or rye, are planted during off-seasons to protect the soil from erosion, suppress weeds, and improve soil fertility. Cover crops add organic matter to the soil and enhance its structure, making it more resilient to erosion and compaction.
- **Crop Rotation:** Rotating crops from season to season helps break pest and disease cycles, improve soil health, and reduce the need for chemical inputs. Different crops have different nutrient requirements and contributions, so rotating them helps balance soil nutrients and prevent depletion.
- **Permanent Soil Cover:** Maintaining a permanent cover on the soil, whether through crop residues, cover crops, or mulch, protects the soil from erosion, reduces water evaporation, and enhances soil organic matter. It also provides habitat for beneficial organisms such as earthworms and microorganisms that improve soil fertility.

## 4. Integrated Pest Management (IPM)

Integrated Pest Management (IPM) is an ecological approach to pest control that combines a variety of management strategies and practices to minimize the use of chemical pesticides. The goal of IPM is to control pest populations to a level that does not cause economic harm while minimizing risks to human health and the environment.

- **Monitoring and Identification:** The first step in IPM is regular monitoring and accurate identification of pests. This allows farmers to make informed decisions about whether pest control is needed and what methods to use.
- **Preventive Practices:** IPM emphasizes preventive practices that reduce the likelihood of pest outbreaks. These include crop rotation, use of resistant varieties, habitat management, and maintaining healthy crops through proper nutrition and irrigation.
- **Biological Control:** Biological control involves using natural predators, parasitoids, or pathogens to control pest populations. For example, ladybugs can be introduced to control aphid populations, and parasitic wasps can be used to target caterpillar pests.
- **Mechanical and Physical Controls:** Mechanical controls include traps, barriers, and hand-picking of pests, while physical controls involve altering the environment to make it less favorable for pests, such as using row covers to protect plants from insects or adjusting planting times to avoid peak pest periods.
- **Chemical Control:** When necessary, IPM allows for the use of chemical pesticides, but they are applied in a targeted and judicious manner. This minimizes the impact on non-target organisms and reduces the risk of pesticide resistance. Often, IPM programs recommend using the least toxic options available.

## 5. Agroforestry

Agroforestry is the practice of integrating trees and shrubs into agricultural landscapes. This approach combines agriculture and forestry to create more diverse, productive, and sustainable land-use systems. Agroforestry has numerous benefits, including enhancing biodiversity, improving soil health, sequestering carbon, and providing additional sources of income for farmers.

- **Silvopasture:** Silvopasture involves integrating trees and shrubs with livestock grazing. Trees provide shade and shelter for animals, improve soil health through leaf litter and root systems, and can produce additional income through timber, fruit, or nuts.

- **Alley Cropping:** Alley cropping involves planting rows of trees or shrubs between rows of crops. The trees can provide windbreaks, reduce soil erosion, and improve nutrient cycling. In some systems, nitrogen-fixing trees are used to enhance soil fertility for the crops grown in the alleys.
- **Forest Farming:** Forest farming is the cultivation of high-value specialty crops such as mushrooms, medicinal plants, or nuts under the canopy of an existing forest. This practice utilizes the shade and microclimate provided by the forest while maintaining forest cover and biodiversity.
- **Windbreaks:** Planting rows of trees or shrubs as windbreaks can protect crops from wind damage, reduce soil erosion, and provide habitat for wildlife. Windbreaks can also contribute to carbon sequestration and improve the microclimate for adjacent crops.

### 6. Water Conservation and Management

Efficient water use is crucial for sustainable farming, particularly in regions where water is scarce. Water conservation practices aim to reduce water usage, improve irrigation efficiency, and protect water resources from pollution and overuse.

- **Efficient Irrigation Systems:** Drip irrigation and sprinkler systems are more efficient than traditional flood irrigation, as they deliver water directly to the plant roots, reducing evaporation and runoff. These systems can be further optimized with soil moisture sensors and automated controls.
- **Rainwater Harvesting:** Collecting and storing rainwater for agricultural use can reduce reliance on external water sources and improve water availability during dry periods. Rainwater harvesting systems can range from simple barrels to more complex catchment and storage systems.
- **Soil Moisture Management:** Practices such as mulching, cover cropping, and no-till farming help retain soil moisture by reducing evaporation and improving soil structure. These practices also enhance the soil's ability to absorb and store water.
- **Water Recycling and Reuse:** Reusing water from agricultural runoff, treated wastewater, or other sources can reduce the demand for fresh water. In some cases, constructed wetlands are used to treat wastewater before it is reused for irrigation.

## Benefits of Sustainable Farming Practices

The adoption of sustainable farming practices offers numerous benefits:

### 1. Environmental Protection

Sustainable farming practices are designed to protect the environment by reducing soil erosion, conserving water, enhancing biodiversity, and minimizing pollution. By maintaining healthy ecosystems, these practices help preserve the natural resources that agriculture depends on.

### 2. Climate Change Mitigation

Sustainable farming practices can play a significant role in mitigating climate change. Practices such as no-till farming, agroforestry, and cover cropping sequester carbon in soils and vegetation, helping to reduce greenhouse gas emissions. Additionally, sustainable farming reduces the need for synthetic inputs, which are often associated with high energy consumption and greenhouse gas emissions.

### 3. Economic Resilience

Sustainable farming practices can improve farm profitability by reducing input costs, increasing productivity, and diversifying income sources. By enhancing the resilience of farming systems to environmental and economic shocks, these practices contribute to long-term economic sustainability.

### 4. Social and Community Benefits

Sustainable farming practices often promote social equity by supporting smallholder farmers, preserving traditional knowledge, and fostering local food systems. They also contribute to food security by increasing the availability of healthy, nutritious food produced in environmentally friendly ways.

## Challenges and Barriers to Adoption

Despite the clear benefits of sustainable farming practices, their adoption is not without challenges. These challenges include:

### 1. Economic Barriers

Transitioning to sustainable farming practices can require significant upfront investments in new equipment, inputs, or infrastructure. For smallholder farmers and those in developing countries, these costs can be prohibitive.

### 2. Knowledge and Skill Gaps

Adopting sustainable farming practices often requires new knowledge and skills. Farmers may need training in techniques such as organic farming, IPM, or water conservation, and access to this training can be limited, particularly in rural areas.

### 3. Policy and Institutional Barriers

In some cases, existing agricultural policies and institutions may not support the adoption of sustainable farming practices. For example, subsidies for chemical inputs or lack of access to credit for sustainable practices can discourage farmers from making the transition.

### 4. Market Access

Sustainable farming practices often produce crops that are sold in niche markets, such as organic or fair-trade markets. Access to these markets can be limited, particularly for smallholder farmers in developing countries.

## Future Prospects and Conclusion

The future of sustainable farming practices is closely tied to the broader goals of global sustainability. As the world faces increasing challenges from climate change, population growth, and resource depletion, the adoption of sustainable farming practices will be essential to ensuring food security and environmental sustainability.

Innovations in technology, policy, and education will play a critical role in overcoming the barriers to adoption and scaling up sustainable farming practices. Continued research and investment in sustainable agriculture will be necessary to develop new techniques and practices that can meet the demands of a growing global population while protecting the environment.

In conclusion, sustainable farming practices offer a pathway to a more resilient, equitable, and sustainable agricultural future. By balancing the needs of food production with the preservation of natural resources and the well-being of communities, these practices can help ensure that agriculture remains viable for future generations. As we move forward, the widespread adoption of sustainable farming practices will be essential in building a food system that is capable of feeding the world while preserving the planet.

## References

Asokan, R., & Murugan, D. (2018). Sustainable agriculture through organic farming in India. Multidisciplinary Global Journal of Academic Research (MGJAR), 5(3).

Dessart, F. J., Barreiro-Hurlé, J., & Van Bavel, R. (2019). Behavioural factors affecting the adoption of sustainable farming practices: a policy-oriented review. European Review of Agricultural Economics, 46(3), 417-471.

Kaur, G. (2013). Sustainable development in agriculture and green farming in India. OIDA International Journal of Sustainable Development, 6(12), 61-64.

Yadav, S. K., Babu, S., Yadav, M. K., Singh, K., Yadav, G. S., & Pal, S. (2013). A review of organic farming for sustainable agriculture in Northern India. International Journal of Agronomy, 2013(1), 718145.

# 3

# Precision Agriculture Technologies with Advancement in GPS, GIS, Drones, Data Analysis

**Abstract**

*Precision agriculture (PA) represents a transformative approach to farming that leverages advanced technologies to optimize crop production, enhance resource management, and improve overall farm efficiency. At the core of PA are advancements in GPS, Geographic Information Systems (GIS), drones, and data analysis. These technologies enable farmers to apply inputs with high precision, monitor field conditions in real-time, and make informed decisions based on comprehensive data. This chapter explores the evolution and impact of precision agriculture technologies, examining how GPS, GIS, drones, and data analytics contribute to more sustainable and efficient farming practices. It also discusses the integration of these technologies, their benefits, challenges, and future prospects in advancing agricultural productivity and sustainability.*

***Keywords:*** *Precision agriculture, GPS, Geographic Information Systems (GIS), drones, data analysis, crop management, farm efficiency, resource optimization, technology integration.*

## Introduction

Precision agriculture (PA) represents a revolutionary shift in farming practices, driven by technological advancements that enable more precise and efficient management of agricultural resources. The integration of Global Positioning Systems (GPS), Geographic Information Systems (GIS), drones, and data analysis has redefined how farmers approach crop management, resource allocation, and field monitoring. By harnessing these technologies, precision agriculture aims to enhance productivity, reduce environmental impact, and optimize the use of inputs such as water, fertilizers, and pesticides.

Precision agriculture technologies, focusing on the advancements in GPS, GIS, drones, and data analysis. It explores how these technologies work together to improve farm operations, the benefits they offer, and the challenges associated with their adoption. Additionally, the chapter examines the future prospects of precision agriculture and its potential to shape the future of farming.

## Evolution of Precision Agriculture Technologies

### Early Developments

The concept of precision agriculture is not new; it has evolved over several decades, driven by advancements in technology and a growing recognition of the need for more efficient farming practices. Early precision agriculture efforts focused on using GPS technology to improve field mapping and yield monitoring.

In the 1990s, GPS technology began to be integrated into agricultural machinery, allowing for more accurate field navigation and data collection. This period marked the beginning of a shift from traditional farming methods to more data-driven approaches.

### The Rise of GIS

The introduction of Geographic Information Systems (GIS) further enhanced precision agriculture by enabling the spatial analysis of agricultural data. GIS technology allowed farmers to overlay various types of data, such as soil properties, crop yields, and weather conditions, onto detailed maps of their fields. This capability provided valuable insights into spatial variability and helped farmers make more informed decisions about input application and resource management.

### Advances in Drones and Remote Sensing

The early 2000s saw significant advancements in drone technology and remote sensing. Drones equipped with high-resolution cameras and sensors became increasingly accessible, providing farmers with a new tool for monitoring crop health, assessing field conditions, and collecting detailed imagery. Remote sensing technology, including satellite imagery, further enhanced the ability to gather data on a large scale, offering insights into crop growth, soil moisture, and other key variables.

### Data Analytics and Big Data

The advent of big data and advanced data analytics has been a game-changer for precision agriculture. The ability to collect, process, and analyze vast amounts of data from various sources has enabled farmers to gain deeper insights into field conditions, optimize input use, and predict crop performance. Data analytics tools and algorithms have become essential for interpreting complex datasets and making data-driven decisions.

## Key Technologies in Precision Agriculture

### 1. Global Positioning Systems (GPS)

Global Positioning Systems (GPS) are foundational to precision agriculture, providing the spatial accuracy needed for precise field operations. GPS technology enables farmers to navigate their fields with high precision, apply inputs with accuracy, and collect spatial data for analysis.

- **Field Mapping and Navigation:** GPS technology allows for the creation of detailed field maps, which can be used for navigation and guidance during planting, spraying, and harvesting. This precision helps reduce overlap and skips, optimizing input use and minimizing waste.
- **Auto-Steering Systems:** GPS-based auto-steering systems help automate machinery operation, improving accuracy and efficiency. These systems ensure that equipment follows predefined paths with minimal deviation, reducing operator fatigue and improving consistency.
- **Variable Rate Technology (VRT):** GPS enables variable rate technology, which adjusts the application rates of inputs such as fertilizers, pesticides, and seeds based on spatial variability. This targeted approach helps optimize resource use and improve crop yields.

### 2. Geographic Information Systems (GIS)

Geographic Information Systems (GIS) provide a powerful platform for analyzing and managing spatial data. In precision agriculture, GIS technology is used to integrate and interpret data from various sources, offering insights into field variability and helping farmers make informed decisions.

- **Spatial Analysis:** GIS allows for the analysis of spatial patterns and variability within fields. By overlaying data such as soil properties, crop yields, and topography, farmers can identify areas with different needs and tailor their management practices accordingly.
- **Decision Support Systems:** GIS-based decision support systems help farmers evaluate different scenarios and make data-driven decisions. These systems can model the impact of various management practices, such as different irrigation strategies or fertilization rates, on crop performance.
- **Data Integration:** GIS integrates data from multiple sources, including GPS, remote sensing, and soil sensors. This integration provides a comprehensive view of field conditions, enabling farmers to make more accurate assessments and plan their operations effectively.

### 3. Drones and Remote Sensing

Drones and remote sensing technologies have revolutionized field monitoring and data collection in precision agriculture. These technologies provide high-resolution imagery and detailed information about crop health, soil conditions, and other key variables.

- **Crop Health Monitoring:** Drones equipped with multispectral and hyperspectral sensors can capture detailed images of crop health, detecting issues such as nutrient deficiencies, pest infestations, and diseases. This information allows for timely intervention and targeted management.
- **Soil and Field Assessment:** Drones can also be used for soil and field assessments, providing data on soil moisture, compaction, and erosion. This information helps farmers make informed decisions about irrigation, tillage, and other field operations.
- **Yield Estimation:** Remote sensing technologies, including satellite imagery and aerial surveys, can be used to estimate crop yields and monitor growth patterns. This data helps farmers plan for harvest and make decisions about marketing and storage.

### 4. Data Analysis and Big Data

The integration of data analytics and big data has transformed precision agriculture by enabling farmers to analyze and interpret large volumes of data. Advanced data analysis tools and algorithms provide valuable insights into field conditions, crop performance, and resource management.

- **Data Integration and Processing:** Data analytics tools can process data from various sources, including GPS, GIS, drones, and sensors. This integration allows for a comprehensive analysis of field conditions and helps identify patterns and trends.
- **Predictive Analytics:** Predictive analytics use historical and real-time data to forecast future conditions and outcomes. In precision agriculture, predictive models can estimate crop yields, forecast weather impacts, and optimize input use based on historical performance and current conditions.
- **Decision-Making Support:** Data analysis tools provide decision-making support by offering actionable insights and recommendations. Farmers can use these tools to evaluate different management strategies, optimize resource allocation, and improve overall farm performance.

## Benefits of Precision Agriculture Technologies

The adoption of precision agriculture technologies offers numerous benefits, including:

### 1. Improved Efficiency

Precision agriculture technologies enhance operational efficiency by reducing overlap and waste. GPS and auto-steering systems improve the accuracy of field operations, while variable rate technology ensures that inputs are applied precisely where needed. This efficiency leads to cost savings and increased productivity.

### 2. Enhanced Resource Management

Precision agriculture allows for more efficient use of resources such as water, fertilizers, and pesticides. By tailoring input application to the specific needs of different areas within a field, farmers can reduce resource use and minimize environmental impact.

### 3. Increased Crop Yields

The ability to monitor and manage field conditions with high precision can lead to increased crop yields. By addressing issues such as nutrient deficiencies, pests, and diseases in a timely manner, farmers can improve crop health and productivity.

### 4. Environmental Sustainability

Precision agriculture technologies contribute to environmental sustainability by reducing the use of chemical inputs, minimizing soil erosion, and conserving water. By optimizing resource use and reducing waste, these technologies help mitigate the environmental impact of farming.

### 5. Data-Driven Decision Making

The integration of data analytics allows farmers to make informed decisions based on comprehensive data. This data-driven approach helps optimize management practices, improve farm performance, and adapt to changing conditions.

## Challenges and Barriers to Adoption

Despite the benefits, the adoption of precision agriculture technologies faces several challenges:

### 1. High Initial Costs

The initial investment in precision agriculture technologies, including GPS equipment, drones, and data analysis tools, can be substantial. For smallholder farmers or those in developing regions, these costs can be a significant barrier to adoption.

### 2. Technical Complexity

Precision agriculture technologies require a certain level of technical expertise and training. Farmers may need to learn how to operate new equipment, interpret data, and integrate various technologies into their operations.

### 3. Data Management and Security

The collection and management of large volumes of data can be challenging. Ensuring data security and privacy is also a concern, as sensitive information about farm operations and performance may be at risk.

### 4. Connectivity and Infrastructure

Access to reliable internet and communication infrastructure is essential for the effective use of precision agriculture technologies. In remote or underserved areas, limited connectivity can hinder the adoption and effectiveness of these technologies.

### 5. Integration and Compatibility

Integrating different technologies and systems can be complex. Ensuring compatibility between GPS systems, GIS software, drones, and data analysis tools is crucial for effective precision agriculture.

## Future Prospects

The future of precision agriculture is promising, with ongoing advancements in technology and increasing adoption across the agricultural sector. Key trends and developments to watch include:

### 1. Advances in Technology

Continued advancements in GPS, GIS, drones, and data analytics will drive further improvements in precision agriculture. Innovations such as autonomous machinery, advanced sensor technologies, and artificial intelligence will enhance the capabilities of precision agriculture and expand its applications.

### 2. Integration of Artificial Intelligence (AI)

Artificial intelligence (AI) and machine learning will play a growing role in precision agriculture. AI algorithms can analyze complex datasets, identify patterns, and make real-time recommendations, further improving decision-making and farm management.

### 3. Expansion of Data Sources

The integration of additional data sources, such as Internet of Things (IoT) sensors and environmental monitoring systems, will provide more comprehensive insights into field conditions and crop performance. This expanded data landscape will enhance the accuracy and effectiveness of precision agriculture.

### 4. Enhanced Connectivity and Infrastructure

Improvements in connectivity and infrastructure, including the expansion of 5G networks and rural broadband access, will facilitate the adoption and use of precision agriculture technologies. Better connectivity will enable real-time data transmission and remote monitoring.

### 5. Increased Adoption and Accessibility

As technology becomes more affordable and accessible, the adoption of precision agriculture will continue to grow. Efforts to reduce the cost of technologies and provide training and support will help more farmers benefit from precision agriculture practices.

## Conclusion

Precision agriculture technologies, including GPS, GIS, drones, and data analysis, have revolutionized modern farming by enabling more precise, efficient, and sustainable agricultural practices. These technologies offer numerous benefits, including improved resource management, increased crop yields, and environmental sustainability. However, challenges such as high initial costs, technical complexity, and data management must be addressed to ensure widespread adoption.

As technology continues to advance and become more accessible, the future of precision agriculture holds great promise. By leveraging innovations in AI, expanding data sources, and improving connectivity, precision agriculture has the potential to transform farming practices and contribute to a more resilient and sustainable agricultural system. The integration of these technologies will be essential in meeting the demands of a growing global population while protecting the environment and optimizing resource use.

## References

Abdullahi, H. S., Mahieddine, F., & Sheriff, R. E. (2015). Technology impact on agricultural productivity: A review of precision agriculture using unmanned aerial vehicles. In Wireless and Satellite Systems: 7th International Conference, WiSATS 2015, Bradford, UK, July 6-7, 2015. Revised Selected Papers 7 (pp. 388-400). Springer International Publishing.

Huang, Y., & Brown, M. (2019). Advancing to the next generation of precision agriculture. Agriculture & food systems to, 2050, 285â.

Mishra, S. (2021). Emerging Technologies—Principles and Applications in Precision Agriculture. Data science in agriculture and natural resource management, 31-53.

Singh, A. K. (2022). Precision agriculture in india–opportunities and challenges. Indian Journal of Fertilisers, 18(4), 308-331.

# 4

# Crop Management and Improvement

**Abstract**

*Crop management and improvement are central to advancing agricultural productivity and sustainability. This comprehensive chapter delves into the strategies and techniques used in modern crop management and the ongoing innovations driving crop improvement. Covering topics such as crop rotation, soil management, pest and disease control, and the application of biotechnology, the chapter provides a detailed analysis of how these practices contribute to higher yields, enhanced quality, and environmental sustainability. The discussion also explores the role of precision agriculture, breeding technologies, and genetic modification in addressing the challenges posed by climate change, resource limitations, and changing market demands. By examining the latest advancements and best practices, this chapter offers valuable insights into the future of crop management and improvement.*

***Keywords:*** *Crop management, crop improvement, crop rotation, soil management, pest control, disease management, biotechnology, precision agriculture, genetic modification, sustainable agriculture.*

## Introduction

Crop management and improvement are pivotal aspects of modern agriculture, playing a critical role in enhancing food security and sustainable agricultural practices. Effective crop management involves optimizing various factors such as soil health, pest control, and crop rotation, while crop improvement focuses on developing higher-yielding, disease-resistant, and climate-adapted crop varieties. As global agricultural challenges evolve, including climate change, resource scarcity, and increasing food demand, the integration of innovative techniques and technologies in crop management and improvement becomes essential.

## Crop Management Practices

### 1. Crop Rotation

Crop rotation is a fundamental crop management practice that involves alternating the types of crops grown in a particular field over successive growing seasons. This practice helps maintain soil fertility, manage pests and diseases, and improve overall crop yield.

- **Soil Fertility Management:** Rotating crops with different nutrient requirements and contributions helps balance soil nutrients. For example, planting legumes, which fix nitrogen in the soil, can enhance soil fertility for subsequent crops.
- **Pest and Disease Control:** Crop rotation disrupts the life cycles of pests and diseases that are specific to certain crops. By alternating crops, farmers can reduce the incidence of pest and disease outbreaks and minimize the need for chemical treatments.
- **Weed Management:** Rotating crops with different growth habits and competitive abilities can help manage weed populations. Different crops can outcompete weeds or require different weed control strategies, reducing weed pressure over time.

### 2. Soil Management

Soil management is crucial for maintaining soil health and ensuring optimal crop growth. Effective soil management practices include soil conservation, nutrient management, and organic matter enhancement.

- **Soil Conservation:** Techniques such as contour plowing, terracing, and reduced tillage help prevent soil erosion and maintain soil structure. These practices protect the topsoil, which is essential for crop growth and productivity.
- **Nutrient Management:** Proper nutrient management involves applying fertilizers and soil amendments based on soil testing and crop requirements. Precision agriculture technologies, such as variable rate fertilization, allow for targeted nutrient application, reducing waste and environmental impact.
- **Organic Matter Enhancement:** Adding organic matter, such as compost or cover crops, improves soil structure, water retention, and microbial activity. Organic matter also contributes to the long-term fertility and health of the soil.

### 3. Pest and Disease Management

Effective pest and disease management is essential for protecting crops and maximizing yield. Integrated Pest Management (IPM) is a comprehensive approach that combines multiple strategies to manage pests and diseases sustainably.

- **Monitoring and Identification:** Regular monitoring and accurate identification of pests and diseases are critical for effective management. Tools such as pheromone traps, visual inspections, and remote sensing technologies can help detect issues early.

- **Preventive Measures:** Preventive measures include practices such as crop rotation, resistant varieties, and sanitation. These strategies help reduce the likelihood of pest and disease outbreaks and minimize the need for chemical interventions.
- **Biological Controls:** Biological controls involve using natural predators, parasites, or pathogens to manage pest populations. For example, releasing ladybugs to control aphids or introducing parasitic wasps to target caterpillar pests.
- **Chemical Controls:** When necessary, targeted application of chemical pesticides can help manage pest and disease outbreaks. IPM emphasizes the use of the least toxic options and applies chemicals in a manner that minimizes harm to non-target organisms and the environment.

### 4. Irrigation Management

Efficient irrigation management is crucial for ensuring adequate water supply and optimizing crop yields. Advances in irrigation technology and practices help conserve water and improve water use efficiency.

- **Irrigation Systems:** Various irrigation systems, such as drip, sprinkler, and furrow irrigation, offer different levels of water efficiency and suitability for different crops and soil types. Drip irrigation, for example, delivers water directly to the plant roots, reducing evaporation and runoff.
- **Water Management:** Effective water management involves scheduling irrigation based on crop needs, weather conditions, and soil moisture levels. Tools such as soil moisture sensors and weather forecasting systems can help optimize irrigation timing and volume.
- **Water Conservation:** Water conservation practices, such as rainwater harvesting and water recycling, help reduce reliance on external water sources and improve water availability during dry periods.

## Crop Improvement Strategies

### 1. Breeding Technologies

Crop breeding is a traditional and ongoing method of crop improvement, aimed at developing new crop varieties with desirable traits such as higher yield, disease resistance, and improved quality.

- **Conventional Breeding:** Conventional breeding involves selecting and crossbreeding plants with desirable traits to develop new varieties. This process may take several generations and relies on natural genetic variation within crop species.

- **Marker-Assisted Selection (MAS):** Marker-assisted selection uses molecular markers to identify and select plants with specific genetic traits. This technique speeds up the breeding process by allowing breeders to select individuals with desirable traits more efficiently.
- **Hybridization:** Hybrid breeding involves crossing two genetically distinct parent lines to produce hybrids with improved traits. Hybrids often exhibit higher yields, better disease resistance, and greater uniformity compared to traditional varieties.

## 2. Genetic Modification

Genetic modification (GM) involves altering the genetic makeup of plants to introduce new traits or enhance existing ones. This technology offers the potential for rapid and precise improvements in crop characteristics.

- **Genetically Modified Crops:** GM crops are engineered to express specific traits, such as resistance to pests, tolerance to herbicides, or improved nutritional content. Examples include Bt cotton, which produces an insecticidal protein to control pests, and Roundup Ready soybeans, which are resistant to glyphosate herbicides.
- **Gene Editing:** Gene editing technologies, such as CRISPR-Cas9, allow for precise modifications to the DNA of plants. Gene editing can be used to enhance traits such as disease resistance, drought tolerance, and nutritional quality without introducing foreign genes.
- **Biofortification:** Biofortification involves enhancing the nutritional content of crops through genetic modification or breeding. This approach aims to address micronutrient deficiencies in populations by increasing the levels of essential nutrients in staple crops.

## 3. Biotechnology and Genomics

Biotechnology and genomics play a significant role in crop improvement by providing tools for understanding and manipulating plant genetics at a molecular level.

- **Genomic Selection:** Genomic selection uses DNA sequencing and molecular markers to predict the performance of plants based on their genetic makeup. This approach helps breeders select individuals with desirable traits more accurately and efficiently.
- **Functional Genomics:** Functional genomics studies the functions and interactions of genes and their products. This knowledge helps identify genes associated with important traits and provides targets for genetic improvement.

- **Proteomics and Metabolomics:** Proteomics and metabolomics involve the analysis of proteins and metabolites in plants. These approaches provide insights into plant physiological responses and help identify biomarkers for crop improvement.

## Advances in Precision Agriculture

Precision agriculture technologies have revolutionized crop management by providing tools for accurate and efficient field operations. Key advancements include:

### 1. Global Positioning Systems (GPS)

Global Positioning Systems (GPS) have revolutionized modern agriculture by providing precise location data that enhances the accuracy and efficiency of various farming operations. GPS technology, originally developed for military applications, has become a fundamental tool in precision agriculture, enabling farmers to optimize their resources, reduce costs, and improve crop yields.

### 2. Precision Farming

One of the primary uses of GPS in agriculture is in precision farming, which involves managing agricultural practices based on detailed spatial information. GPS allows farmers to create accurate maps of their fields, identifying variations in soil properties, crop health, and other critical factors. These maps can then be used to apply inputs such as fertilizers, seeds, and pesticides more precisely, ensuring that each area of the field receives the optimal treatment.

- **Variable Rate Technology (VRT):** GPS is integral to VRT, where input application rates are adjusted based on location-specific data. For example, a GPS-guided tractor can vary the amount of fertilizer it applies as it moves across different zones of a field, reducing waste and improving crop performance.
- **Field Mapping:** With GPS, farmers can create detailed maps of their fields, including boundaries, irrigation systems, and areas with specific soil types or crop conditions. This information helps in planning and monitoring field operations more effectively.

### 3. Automated Machinery and Guidance Systems

GPS technology has enabled the development of automated machinery and guidance systems, which are increasingly common in modern agriculture. These systems allow tractors, harvesters, and other equipment to operate with minimal human intervention, following precise paths across fields.

- **Auto-Steering:** GPS-guided auto-steering systems enable tractors and other machinery to follow exact routes within a field, reducing overlap and ensuring that all areas are covered uniformly. This increases efficiency, reduces fuel consumption, and minimizes soil compaction.

- **Reduced Operator Fatigue:** By automating many of the repetitive tasks involved in fieldwork, GPS-guided machinery reduces the physical and mental strain on operators, allowing them to focus on more complex aspects of farm management.
- **Night and Low-Visibility Operations:** GPS enables farming operations to continue during low-visibility conditions, such as at night or in foggy weather, by guiding machinery accurately even when the operator's visibility is limited.

### 4. Yield Monitoring and Mapping

GPS is also essential for yield monitoring and mapping, which involves collecting and analyzing data on crop yields as they are harvested. This information helps farmers understand how different areas of a field are performing and make informed decisions about future planting and management practices.

- **Yield Maps:** By combining GPS data with yield sensors on harvesting equipment, farmers can generate yield maps that show the productivity of different field zones. These maps are crucial for identifying areas that may require additional attention, such as soil amendments or drainage improvements.
- **Data-Driven Decisions:** The insights gained from yield monitoring and mapping enable farmers to make data-driven decisions, such as adjusting planting density, selecting crop varieties, or modifying input applications to improve overall productivity.

### 5. Soil Sampling and Analysis

GPS plays a crucial role in soil sampling and analysis by ensuring that samples are taken from precise locations within a field. This allows for more accurate assessment of soil properties, such as pH, nutrient levels, and organic matter content.

- **Grid Sampling:** Farmers often use a grid sampling approach, where soil samples are collected at regular intervals across a field. GPS ensures that each sample is taken from the correct location, making it possible to create detailed soil maps that guide targeted soil management practices.
- **Site-Specific Management:** With precise soil data, farmers can implement site-specific management practices, such as variable rate fertilization or lime application, to address the unique needs of different areas within a field.

## 6. Irrigation Management

GPS is instrumental in improving irrigation management by helping farmers design and implement more efficient irrigation systems. Accurate field mapping with GPS enables the creation of tailored irrigation plans that optimize water use and reduce waste.

- **Precision Irrigation:** GPS-guided systems can control the timing and amount of water applied to specific areas of a field, ensuring that crops receive the right amount of water when and where it is needed. This reduces water consumption, lowers costs, and helps conserve water resources.
- **Drainage Mapping:** GPS can also be used to map and manage field drainage systems, identifying areas prone to waterlogging and guiding the installation of drainage infrastructure to improve soil conditions and prevent crop loss.

## 7. Livestock Management

In addition to crop management, GPS technology is increasingly used in livestock management, providing farmers with tools to monitor and manage their herds more effectively.

- **Animal Tracking:** GPS collars on livestock allow farmers to track the location and movement of their animals in real-time. This helps in managing grazing patterns, preventing overgrazing, and ensuring that livestock remain within designated areas.
- **Health Monitoring:** By tracking animal movement and behavior, GPS systems can provide early warnings of potential health issues, such as lameness or illness, allowing for prompt intervention.

## 8. Environmental Monitoring

GPS is also a valuable tool for environmental monitoring, helping farmers to assess and manage the impact of their farming practices on the surrounding environment.

- **Erosion Control:** GPS mapping can identify areas at risk of soil erosion, guiding the implementation of erosion control measures such as contour farming, terracing, or the planting of cover crops.
- **Biodiversity Management:** Farmers can use GPS to monitor and manage biodiversity on their farms, such as mapping and protecting habitats for beneficial insects or creating buffer zones around waterways to reduce runoff and protect water quality.

## 2. Geographic Information Systems (GIS)

Geographic Information Systems (GIS) are powerful tools that have significantly transformed the agricultural sector by enabling the collection, analysis, and visualization of spatial data. GIS technology allows farmers, agronomists, and researchers to understand and manage the complexities of agricultural landscapes with unprecedented accuracy. The integration of GIS in agriculture supports a wide range of applications, from precision farming to resource management, ultimately contributing to more efficient, productive, and sustainable farming practices.

### 1. Spatial Data Collection and Mapping

At the core of GIS is its ability to collect and manage spatial data—information that has a geographic or locational component. In agriculture, GIS is used to create detailed maps that represent various aspects of a farm or agricultural landscape.

- **Field Mapping:** GIS enables the creation of accurate maps of farm fields, including boundaries, soil types, elevation, and land use patterns. These maps serve as the foundation for precision agriculture practices by allowing farmers to visualize the spatial variability within their fields.
- **Soil and Crop Mapping:** Using data from soil samples, remote sensing, and on-ground sensors, GIS can produce detailed maps showing soil properties (e.g., pH, nutrient levels, organic matter) and crop health. These maps help farmers identify areas with specific needs, such as nutrient deficiencies or pest problems, and manage them accordingly.
- **Topography and Drainage Mapping:** GIS allows for the creation of topographic maps that show the elevation and slope of fields. This information is critical for planning irrigation and drainage systems, as it helps farmers design systems that optimize water distribution and prevent soil erosion.

### 2. Precision Agriculture and Site-Specific Management

GIS plays a central role in precision agriculture by enabling site-specific management practices. These practices involve tailoring farming inputs—such as seeds, fertilizers, and water—to the specific conditions of different areas within a field.

- **Variable Rate Application (VRA):** By integrating GIS with GPS and other technologies, farmers can implement VRA, where inputs are applied at varying rates across a field based on spatial data. For example, GIS can help determine which areas of a field require more or less fertilizer, ensuring that resources are used efficiently and crop yields are maximized.

- **Zoning and Field Segmentation:** GIS allows fields to be divided into management zones based on characteristics such as soil type, crop performance, or moisture levels. These zones can be managed independently, enabling farmers to optimize inputs and practices for each specific area.
- **Yield Monitoring and Analysis:** GIS is used to map and analyze crop yields across different parts of a field. By correlating yield data with other spatial data (e.g., soil properties, topography), farmers can identify factors that influence productivity and make informed decisions to improve future yields.

### 3. Environmental and Resource Management

GIS is a valuable tool for managing natural resources and mitigating the environmental impact of agricultural practices. It helps farmers and land managers make data-driven decisions that balance productivity with sustainability.

- **Water Management:** GIS supports the design and management of efficient irrigation systems by mapping water sources, soil moisture levels, and crop water requirements. It also aids in the monitoring of water use, helping to prevent over-irrigation and reduce water waste.
- **Soil Conservation:** GIS can identify areas at risk of soil erosion, enabling the implementation of conservation practices such as contour plowing, terracing, and the establishment of cover crops. By mapping erosion-prone areas, farmers can target their efforts to maintain soil health and prevent land degradation.
- **Biodiversity and Habitat Protection:** GIS helps in mapping and protecting areas of high biodiversity value on farms, such as wetlands, riparian zones, and wildlife corridors. Farmers can use this information to implement conservation practices that enhance biodiversity while maintaining productive agricultural land.

### 4. Crop Monitoring and Remote Sensing

GIS, when integrated with remote sensing technologies such as satellites and drones, enables real-time monitoring of crop conditions and environmental factors. This combination provides farmers with timely and accurate information to make proactive management decisions.

- **Vegetation Index Mapping:** Remote sensing data, processed through GIS, can be used to create vegetation index maps (e.g., NDVI—Normalized Difference Vegetation Index) that indicate crop health and vigor. These maps help farmers identify areas of stress due to pests, diseases, or nutrient deficiencies, allowing for targeted interventions.

- **Pest and Disease Monitoring:** GIS can map the spread of pests and diseases across a landscape, helping farmers and agricultural advisors to track outbreaks and plan control measures. This spatial information is crucial for implementing integrated pest management (IPM) strategies that minimize chemical use and environmental impact.
- **Weather and Climate Monitoring:** GIS integrates weather data with spatial information to assess how climate conditions affect crop growth and development. Farmers can use this information to adjust planting schedules, irrigation plans, and other management practices in response to changing weather patterns.

## 5. Planning and Decision Support Systems

GIS is increasingly used in the development of decision support systems (DSS) for agriculture, providing farmers and land managers with tools to plan and optimize their operations.

- **Farm Planning and Layout:** GIS helps farmers design the layout of their farms, including the placement of fields, irrigation systems, and infrastructure. By analyzing spatial data, farmers can optimize land use and improve operational efficiency.
- **Crop Rotation and Land Use Planning:** GIS supports the planning of crop rotations by mapping soil fertility, crop performance, and other factors. This helps farmers design rotations that enhance soil health, reduce pest pressures, and improve overall farm sustainability.
- **Risk Assessment and Management:** GIS can model and analyze risks such as drought, flooding, or pest outbreaks. This information helps farmers develop risk management strategies, such as diversifying crops, adjusting planting dates, or investing in infrastructure to mitigate potential impacts.

## 6. Supply Chain and Market Analysis

GIS is also valuable beyond the farm gate, supporting agricultural supply chains and market analysis by mapping the spatial distribution of production, processing, and distribution facilities.

- **Logistics and Transportation Planning:** GIS can optimize the transportation of agricultural products by mapping the locations of farms, processing plants, and markets. This helps reduce transportation costs, minimize delays, and ensure that products reach markets in the best condition.
- **Market Analysis and Forecasting:** GIS can analyze the spatial distribution of agricultural production and consumption, helping farmers

and agribusinesses identify potential markets and forecast demand. This spatial analysis supports strategic decisions about crop selection, production levels, and marketing strategies.

- **Traceability and Certification:** GIS supports traceability in agricultural supply chains by mapping the origins of products and tracking their movement through the supply chain. This is increasingly important for meeting consumer demands for transparency and for complying with food safety and sustainability certifications.

### 7. Climate Change Adaptation

As climate change continues to impact agriculture, GIS is becoming an essential tool for helping farmers adapt to changing conditions and mitigate risks.

- **Climate Impact Assessment:** GIS can model the potential impacts of climate change on agriculture, such as changes in temperature, precipitation, and growing seasons. This information helps farmers plan for future conditions and develop strategies to cope with climate variability.
- **Adaptation Planning:** GIS supports the development of adaptation strategies, such as selecting climate-resilient crops, redesigning irrigation systems, or altering land use patterns. By analyzing spatial data, farmers can identify the most vulnerable areas and implement targeted measures to enhance resilience.
- **Carbon Sequestration and Greenhouse Gas Monitoring:** GIS can map and monitor practices that sequester carbon, such as agroforestry or conservation tillage. This helps farmers contribute to climate change mitigation by reducing their carbon footprint and participating in carbon markets.

## 3. Drones and Remote Sensing

Drones and remote sensing technologies have revolutionized agriculture by providing farmers with precise, real-time data that enhances crop management, improves resource efficiency, and increases productivity. These technologies allow for the monitoring of large areas of farmland quickly and accurately, helping to identify problems early and enabling targeted interventions. The integration of drones with other precision agriculture tools, such as Geographic Information Systems (GIS) and Global Positioning Systems (GPS), has further expanded their capabilities, making them essential tools in modern farming.

## 1. Overview of Drone Technology in Agriculture

Drones, also known as Unmanned Aerial Vehicles (UAVs), are equipped with various sensors and cameras that capture detailed images and data from above. These aerial platforms can cover vast areas of land in a short amount of time, providing high-resolution images that can be analyzed to assess crop health, soil conditions, and other critical factors.

- **Types of Drones:** Agricultural drones vary in size, capability, and sensor types. Fixed-wing drones are often used for covering large areas due to their extended flight times, while multi-rotor drones offer greater maneuverability and are ideal for more detailed inspections of specific areas.
- **Sensors and Cameras:** Drones can be equipped with a variety of sensors, including RGB cameras, multispectral cameras, thermal cameras, and LiDAR (Light Detection and Ranging) sensors. These sensors capture different types of data, from visible light images to detailed measurements of vegetation health and soil moisture.

## 2. Crop Monitoring and Health Assessment

One of the primary uses of drones in agriculture is for crop monitoring and health assessment. By capturing high-resolution images and data, drones provide farmers with insights into the condition of their crops, allowing for early detection of issues and informed decision-making.

- **Vegetation Indices:** Drones equipped with multispectral cameras can capture images that are processed to calculate vegetation indices, such as the Normalized Difference Vegetation Index (NDVI). NDVI is a key indicator of plant health, showing how much light is being absorbed by the vegetation. This helps in identifying areas of stress due to nutrient deficiencies, water stress, or pest infestations.
- **Disease Detection:** Drones can detect early signs of crop diseases by capturing images in wavelengths that are not visible to the naked eye. For example, thermal cameras can identify areas of the field where plants are experiencing heat stress, which can be an early indicator of disease or pest problems.
- **Yield Estimation:** By analyzing the health and density of crops throughout the growing season, drones can help farmers estimate yields more accurately. This information is valuable for planning harvests and managing supply chains.

### 3. Precision Agriculture and Variable Rate Application

Drones play a crucial role in precision agriculture by enabling site-specific management practices. The data collected by drones can be used to create detailed maps that guide variable rate applications of inputs such as fertilizers, pesticides, and water.

- **Prescription Maps:** Using drone data, farmers can create prescription maps that define specific rates of input application for different zones within a field. For example, areas identified as having nutrient deficiencies can be targeted with higher rates of fertilizer, while healthier areas receive less, optimizing input use and reducing waste.
- **Targeted Spraying:** Some drones are equipped with spraying systems that can apply pesticides or fertilizers directly to specific areas of the field. This targeted approach reduces chemical use, minimizes environmental impact, and ensures that only the affected areas are treated.
- **Irrigation Management:** Drones can monitor soil moisture levels and crop water needs, helping farmers optimize irrigation practices. By identifying areas of the field that are under- or over-watered, farmers can adjust irrigation schedules and systems to conserve water and improve crop performance.

### 4. Soil Analysis and Field Mapping

Drones are also valuable tools for soil analysis and field mapping, providing detailed information on soil conditions and topography that is essential for effective farm management.

- **Topographic Mapping:** Drones equipped with LiDAR or photogrammetry sensors can create high-resolution topographic maps of fields. These maps show the elevation and slope of the land, helping farmers design efficient irrigation and drainage systems and prevent soil erosion.
- **Soil Condition Monitoring:** Drones can assess soil conditions by capturing data on soil moisture, temperature, and organic matter content. This information helps farmers manage soil health, plan planting schedules, and optimize input applications.
- **Field Boundary Mapping:** Drones can accurately map field boundaries, helping farmers document land use and plan field operations. This is particularly useful for large or irregularly shaped fields where traditional surveying methods may be less efficient.

## 5. Pest and Weed Management

Drones are increasingly being used for pest and weed management, offering a precise and efficient way to monitor and control these threats to crop production.

- **Pest Detection:** By capturing high-resolution images, drones can detect early signs of pest infestations, such as changes in plant color or texture that indicate the presence of pests. Early detection allows for timely intervention, reducing the spread of pests and minimizing crop damage.
- **Weed Identification:** Drones equipped with multispectral or hyperspectral cameras can differentiate between crops and weeds based on their spectral signatures. This allows farmers to identify weed infestations and apply herbicides only where needed, reducing chemical use and protecting crop health.
- **Biological Control:** Drones can also be used to release beneficial insects, such as ladybugs or parasitic wasps, directly into infested areas. This targeted biological control method reduces the need for chemical pesticides and supports integrated pest management (IPM) strategies.

## 6. Harvest Planning and Management

Drones play a significant role in harvest planning and management by providing real-time data that helps farmers make informed decisions about when and how to harvest their crops.

- **Maturity Assessment:** Drones can monitor crop maturity by analyzing color changes and other indicators of ripeness. This information helps farmers determine the optimal time to harvest, ensuring that crops are harvested at their peak quality.
- **Harvest Efficiency:** Drones can map and monitor the progress of harvest operations, identifying areas that have been harvested and those that still need attention. This improves the efficiency of harvest crews and equipment, reducing downtime and ensuring timely completion of harvests.
- **Post-Harvest Monitoring:** After the harvest, drones can be used to assess the condition of the fields, identifying any areas of crop residue or soil compaction that may require management before the next planting season.

## 7. Environmental Monitoring and Conservation

Drones contribute to environmental monitoring and conservation efforts by providing detailed data on land use, vegetation cover, and ecosystem health.

- **Habitat Monitoring:** Drones can monitor and map natural habitats, such as wetlands or riparian zones, that are located within or near agricultural fields. This helps farmers implement conservation practices that protect biodiversity and maintain ecosystem services.

- **Erosion Control:** By monitoring changes in land cover and topography, drones can identify areas at risk of soil erosion. Farmers can use this information to implement erosion control measures, such as planting cover crops or constructing terraces, to protect soil health.
- **Carbon Sequestration:** Drones can monitor practices that enhance carbon sequestration, such as agroforestry or conservation tillage. By capturing data on vegetation growth and soil conditions, drones help farmers measure the impact of these practices on carbon storage and climate change mitigation.

### 8. Research and Development

Drones are also valuable tools for agricultural research and development, providing researchers with detailed data that supports the study of crop genetics, breeding, and agronomy.

- **Field Trials:** Drones can monitor and document the performance of different crop varieties or management practices in field trials. This data is essential for evaluating the effectiveness of new technologies or practices and for developing recommendations for farmers.
- **Genetic Research:** By capturing data on plant phenotypes, drones support genetic research aimed at developing new crop varieties with improved traits, such as disease resistance, drought tolerance, or higher yields.
- **Precision Experimentation:** Drones enable precise experimentation by allowing researchers to apply treatments, monitor results, and analyze data with a high degree of accuracy. This supports the development of precision agriculture practices that can be tailored to specific crops, soils, and climates.

### 4. Data Analytics and Big Data

Drones and remote sensing technologies have revolutionized agriculture by providing farmers with precise, real-time data that enhances crop management, improves resource efficiency, and increases productivity. These technologies allow for the monitoring of large areas of farmland quickly and accurately, helping to identify problems early and enabling targeted interventions. The integration of drones with other precision agriculture tools, such as Geographic Information Systems (GIS) and Global Positioning Systems (GPS), has further expanded their capabilities, making them essential tools in modern farming.

## 1. Overview of Drone Technology in Agriculture

Drones, also known as Unmanned Aerial Vehicles (UAVs), are equipped with various sensors and cameras that capture detailed images and data from above. These aerial platforms can cover vast areas of land in a short amount of time, providing high-resolution images that can be analyzed to assess crop health, soil conditions, and other critical factors.

- **Types of Drones:** Agricultural drones vary in size, capability, and sensor types. Fixed-wing drones are often used for covering large areas due to their extended flight times, while multi-rotor drones offer greater maneuverability and are ideal for more detailed inspections of specific areas.
- **Sensors and Cameras:** Drones can be equipped with a variety of sensors, including RGB cameras, multispectral cameras, thermal cameras, and LiDAR (Light Detection and Ranging) sensors. These sensors capture different types of data, from visible light images to detailed measurements of vegetation health and soil moisture.

## 2. Crop Monitoring and Health Assessment

One of the primary uses of drones in agriculture is for crop monitoring and health assessment. By capturing high-resolution images and data, drones provide farmers with insights into the condition of their crops, allowing for early detection of issues and informed decision-making.

- **Vegetation Indices:** Drones equipped with multispectral cameras can capture images that are processed to calculate vegetation indices, such as the Normalized Difference Vegetation Index (NDVI). NDVI is a key indicator of plant health, showing how much light is being absorbed by the vegetation. This helps in identifying areas of stress due to nutrient deficiencies, water stress, or pest infestations.
- **Disease Detection:** Drones can detect early signs of crop diseases by capturing images in wavelengths that are not visible to the naked eye. For example, thermal cameras can identify areas of the field where plants are experiencing heat stress, which can be an early indicator of disease or pest problems.
- **Yield Estimation:** By analyzing the health and density of crops throughout the growing season, drones can help farmers estimate yields more accurately. This information is valuable for planning harvests and managing supply chains.

### 3. Precision Agriculture and Variable Rate Application

Drones play a crucial role in precision agriculture by enabling site-specific management practices. The data collected by drones can be used to create detailed maps that guide variable rate applications of inputs such as fertilizers, pesticides, and water.

- **Prescription Maps:** Using drone data, farmers can create prescription maps that define specific rates of input application for different zones within a field. For example, areas identified as having nutrient deficiencies can be targeted with higher rates of fertilizer, while healthier areas receive less, optimizing input use and reducing waste.
- **Targeted Spraying:** Some drones are equipped with spraying systems that can apply pesticides or fertilizers directly to specific areas of the field. This targeted approach reduces chemical use, minimizes environmental impact, and ensures that only the affected areas are treated.
- **Irrigation Management:** Drones can monitor soil moisture levels and crop water needs, helping farmers optimize irrigation practices. By identifying areas of the field that are under- or over-watered, farmers can adjust irrigation schedules and systems to conserve water and improve crop performance.

### 4. Soil Analysis and Field Mapping

Drones are also valuable tools for soil analysis and field mapping, providing detailed information on soil conditions and topography that is essential for effective farm management.

- **Topographic Mapping:** Drones equipped with LiDAR or photogrammetry sensors can create high-resolution topographic maps of fields. These maps show the elevation and slope of the land, helping farmers design efficient irrigation and drainage systems and prevent soil erosion.
- **Soil Condition Monitoring:** Drones can assess soil conditions by capturing data on soil moisture, temperature, and organic matter content. This information helps farmers manage soil health, plan planting schedules, and optimize input applications.
- **Field Boundary Mapping:** Drones can accurately map field boundaries, helping farmers document land use and plan field operations. This is particularly useful for large or irregularly shaped fields where traditional surveying methods may be less efficient.

## 5. Pest and Weed Management

Drones are increasingly being used for pest and weed management, offering a precise and efficient way to monitor and control these threats to crop production.

- **Pest Detection:** By capturing high-resolution images, drones can detect early signs of pest infestations, such as changes in plant color or texture that indicate the presence of pests. Early detection allows for timely intervention, reducing the spread of pests and minimizing crop damage.
- **Weed Identification:** Drones equipped with multispectral or hyperspectral cameras can differentiate between crops and weeds based on their spectral signatures. This allows farmers to identify weed infestations and apply herbicides only where needed, reducing chemical use and protecting crop health.
- **Biological Control:** Drones can also be used to release beneficial insects, such as ladybugs or parasitic wasps, directly into infested areas. This targeted biological control method reduces the need for chemical pesticides and supports integrated pest management (IPM) strategies.

## 6. Harvest Planning and Management

Drones play a significant role in harvest planning and management by providing real-time data that helps farmers make informed decisions about when and how to harvest their crops.

- **Maturity Assessment:** Drones can monitor crop maturity by analyzing color changes and other indicators of ripeness. This information helps farmers determine the optimal time to harvest, ensuring that crops are harvested at their peak quality.
- **Harvest Efficiency:** Drones can map and monitor the progress of harvest operations, identifying areas that have been harvested and those that still need attention. This improves the efficiency of harvest crews and equipment, reducing downtime and ensuring timely completion of harvests.
- **Post-Harvest Monitoring:** After the harvest, drones can be used to assess the condition of the fields, identifying any areas of crop residue or soil compaction that may require management before the next planting season.

## 7. Environmental Monitoring and Conservation

Drones contribute to environmental monitoring and conservation efforts by providing detailed data on land use, vegetation cover, and ecosystem health.

- **Habitat Monitoring:** Drones can monitor and map natural habitats, such as wetlands or riparian zones, that are located within or near agricultural fields. This helps farmers implement conservation practices that protect biodiversity and maintain ecosystem services.

- **Erosion Control:** By monitoring changes in land cover and topography, drones can identify areas at risk of soil erosion. Farmers can use this information to implement erosion control measures, such as planting cover crops or constructing terraces, to protect soil health.
- **Carbon Sequestration:** Drones can monitor practices that enhance carbon sequestration, such as agroforestry or conservation tillage. By capturing data on vegetation growth and soil conditions, drones help farmers measure the impact of these practices on carbon storage and climate change mitigation.

### 8. Research and Development

Drones are also valuable tools for agricultural research and development, providing researchers with detailed data that supports the study of crop genetics, breeding, and agronomy.

- **Field Trials:** Drones can monitor and document the performance of different crop varieties or management practices in field trials. This data is essential for evaluating the effectiveness of new technologies or practices and for developing recommendations for farmers.
- **Genetic Research:** By capturing data on plant phenotypes, drones support genetic research aimed at developing new crop varieties with improved traits, such as disease resistance, drought tolerance, or higher yields.
- **Precision Experimentation:** Drones enable precise experimentation by allowing researchers to apply treatments, monitor results, and analyze data with a high degree of accuracy. This supports the development of precision agriculture practices that can be tailored to specific crops, soils, and climates.

## Challenges and Future Directions

While crop management and improvement strategies offer numerous benefits, they also face challenges:

### 1. Climate Change

Climate change impacts crop growth and productivity through altered temperature, precipitation patterns, and increased frequency of extreme weather events. Adaptation strategies, including developing climate-resilient crops and modifying management practices, are essential to address these challenges.

### 2. Resource Limitations

Limited resources, such as water and soil nutrients, pose challenges for crop management. Sustainable practices, including efficient resource use and conservation techniques, are crucial for addressing resource constraints and ensuring long-term productivity.

### 3. Technological Adoption

The adoption of advanced technologies in crop management and improvement can be limited by factors such as cost, technical expertise, and infrastructure. Efforts to reduce costs, provide training, and improve access to technology are necessary to support widespread adoption.

### 4. Market Demands

Changing market demands and consumer preferences influence crop management and improvement strategies. Farmers and breeders must adapt to evolving market trends, including increasing demand for organic and specialty crops, and address issues related to crop quality and nutritional value.

Crop management and improvement are critical components of modern agriculture, driving advances in productivity, sustainability, and resilience. Effective crop management practices, including crop rotation, soil management, pest and disease control, and irrigation management, contribute to optimal crop growth and yield. Meanwhile, crop improvement strategies, such as breeding technologies, genetic modification, and biotechnology, enhance crop traits and address emerging challenges.

The integration of precision agriculture technologies, including GPS, GIS, drones, and data analytics, further enhances crop management by providing tools for accurate and efficient field operations. As agricultural challenges evolve, including climate change and resource limitations, continued innovation and adaptation in crop management and improvement will be essential for meeting global food demands and promoting sustainable agriculture.

By leveraging advancements in technology and adopting best practices, farmers and researchers can drive the future of crop management and improvement, ensuring a resilient and productive agricultural system for generations to come.

## Challenges and Future Directions in Crop Management and Improvement

As agriculture continues to evolve in response to global demands and environmental pressures, several challenges and future directions emerge in the fields of crop management and improvement. Addressing these challenges and exploring future opportunities are crucial for ensuring the sustainability, productivity, and resilience of agricultural systems.

### 1. Climate Change

**Challenges:**

- **Temperature Fluctuations:** Climate change leads to increased temperatures and more frequent heatwaves, which can negatively affect crop growth, development, and yields. High temperatures can reduce photosynthesis, alter flowering times, and increase the risk of heat stress in crops.
- **Altered Precipitation Patterns:** Changes in precipitation patterns, including increased frequency of droughts and heavy rainfall events, can impact soil moisture levels and crop water availability. This variability can lead to reduced crop yields and increased risk of soil erosion and nutrient leaching.
- **Increased Pest and Disease Pressure:** Climate change can influence the distribution and lifecycle of pests and diseases, leading to increased incidence and severity. Warmer temperatures and altered humidity levels can create more favorable conditions for pest and disease proliferation.

**Future Directions**

- **Climate-Resilient Crops:** Developing crop varieties that are more tolerant to extreme temperatures, drought, and variable rainfall is essential. This includes breeding for traits such as heat and drought resistance, as well as improving water use efficiency.
- **Adaptation Strategies:** Implementing adaptive management practices, such as adjusting planting dates, altering irrigation strategies, and utilizing soil conservation techniques, can help mitigate the impacts of climate change.
- **Integrated Approaches:** Combining breeding, agronomic practices, and precision agriculture technologies can enhance resilience. For example, integrating climate data with precision irrigation systems can optimize water use and reduce the impact of variable rainfall.

### 2. Resource Limitations

**Challenges**

- **Water Scarcity:** Increasing competition for water resources and changing precipitation patterns contribute to water scarcity, affecting irrigation and crop yields. Inefficient water use can exacerbate the problem, leading to over-extraction and environmental degradation.
- **Soil Degradation:** Intensive farming practices, including monoculture and excessive tillage, can lead to soil degradation, including erosion,

loss of organic matter, and reduced fertility. Soil degradation impacts crop productivity and long-term sustainability.

- **Nutrient Management:** Ensuring the balanced and efficient use of fertilizers is critical to avoid nutrient runoff and pollution. Over-reliance on chemical inputs can lead to environmental issues such as eutrophication and soil acidification.

**Future Directions**

- **Water-Efficient Technologies:** Advancing irrigation technologies, such as drip and precision irrigation systems, can improve water use efficiency. Implementing water-saving practices, including rainwater harvesting and soil moisture management, can also help address water scarcity.
- **Soil Health Management:** Adopting conservation tillage, cover cropping, and organic matter addition can improve soil health and prevent degradation. These practices enhance soil structure, water retention, and nutrient availability.
- **Sustainable Nutrient Management:** Utilizing precision agriculture tools to apply fertilizers based on soil and crop needs can reduce nutrient loss and environmental impact. Integrated nutrient management approaches that combine organic and synthetic sources can also enhance soil fertility.

**3. Technological Adoption**

**Challenges:**

- **High Costs:** The initial investment required for advanced technologies, such as precision agriculture equipment, genetic modification tools, and data analytics software, can be prohibitive for smallholder farmers and those in developing regions.
- **Technical Expertise:** The complexity of modern technologies requires specialized knowledge and skills. Farmers may need training and support to effectively use new tools and interpret data for informed decision-making.
- **Infrastructure Limitations:** Inadequate infrastructure, including internet connectivity and access to technology, can limit the adoption and effectiveness of precision agriculture technologies in remote or underserved areas.

### Future Directions

- **Cost Reduction:** Efforts to reduce the cost of technologies through innovation, subsidies, or scaled production can increase accessibility for a broader range of farmers. Affordable and user-friendly solutions can help bridge the gap between large-scale and smallholder operations.
- **Education and Training:** Providing training programs and resources to build technical expertise is essential for successful technology adoption. Collaborations between research institutions, extension services, and technology providers can support knowledge transfer and capacity building.
- **Infrastructure Development:** Investing in infrastructure, such as rural broadband and technology access points, can enhance connectivity and support the implementation of precision agriculture technologies. Public-private partnerships may play a role in addressing infrastructure gaps.

## 4. Market Demands and Consumer Preferences

### Challenges

- **Changing Market Demands:** Consumer preferences are evolving towards organic, non-GMO, and specialty crops. Farmers may face challenges in adapting to these changing demands, including adjusting production practices and meeting certification requirements.
- **Global Competition:** Increased competition in global markets requires farmers to remain competitive by improving crop quality, yield, and cost-efficiency. Meeting international standards and navigating complex trade regulations can be challenging.
- **Value Chain Integration:** Integrating with value chains, including processing, distribution, and retail, is crucial for maximizing market opportunities. However, farmers may face difficulties in establishing and managing these connections.

### Future Directions

- **Diversification and Innovation:** Developing and marketing niche crops and value-added products can help meet changing consumer demands. Innovation in crop breeding and production practices can provide new market opportunities and enhance competitiveness.
- **Sustainability Certifications:** Adopting sustainable practices and obtaining certifications, such as organic or fair trade, can help

farmers access premium markets and meet consumer preferences for environmentally friendly products.

- **Supply Chain Collaboration:** Strengthening relationships with stakeholders across the supply chain, including processors, distributors, and retailers, can improve market access and value chain integration. Collaborative approaches can enhance efficiency and create new opportunities.

## Conclusion

Crop management and improvement are critical components of modern agriculture, driving advances in productivity, sustainability, and resilience. Effective crop management practices, including crop rotation, soil management, pest and disease control, and irrigation management, contribute to optimal crop growth and yield. Meanwhile, crop improvement strategies, such as breeding technologies, genetic modification, and biotechnology, enhance crop traits and address emerging challenges.

The integration of precision agriculture technologies, including GPS, GIS, drones, and data analytics, further enhances crop management by providing tools for accurate and efficient field operations. As agricultural challenges evolve, including climate change and resource limitations, continued innovation and adaptation in crop management and improvement will be essential for meeting global food demands and promoting sustainable agriculture.

By leveraging advancements in technology and adopting best practices, farmers and researchers can drive the future of crop management and improvement, ensuring a resilient and productive agricultural system for generations to come.

## References

Pooniya, V., Choudhary, A. K., Dass, A., Bana, R. S., Rana, K. S., Rana, D. S., ... & Puniya, M. M. (2015). Improved crop management practices for sustainable pulse production: An Indian perspective. The Indian Journal of Agricultural Sciences, 85(6), 747-458.

Reddy, B. V. S., Reddy, P. S., Bidinger, F., & Blümmel, M. (2003). Crop management factors influencing yield and quality of crop residues. Field Crops Research, 84(1-2), 57-77.

Shah, F., & Wu, W. (2019). Soil and crop management strategies to ensure higher crop productivity within sustainable environments. Sustainability, 11(5), 1485.

Zuo, Y., & Zhang, F. (2011). Soil and crop management strategies to prevent iron deficiency in crops. Plant and Soil, 339, 83-95.

# 5

# Livestock Management and Welfare

**Abstract**

*Livestock management and welfare are integral to the success of modern animal agriculture and have profound implications for productivity, animal health, and ethical considerations. This chapter explores comprehensive strategies for effective livestock management, including nutrition, housing, breeding, and health care. It also delves into welfare practices that ensure humane treatment and enhance the quality of life for animals. The discussion extends to contemporary issues such as environmental impacts, regulatory frameworks, and innovations in animal care. By examining best practices and emerging trends, this chapter aims to provide a detailed understanding of how to balance productivity with ethical considerations in livestock farming.*

***Keywords:*** *Livestock management, animal welfare, nutrition, housing, breeding, health care, ethical treatment, environmental impact, regulatory frameworks, innovations in animal care.*

## Introduction

Livestock management and welfare are critical aspects of animal agriculture that impact the efficiency, productivity, and ethical standards of farming practices. Effective livestock management encompasses a range of activities, from optimizing nutrition and breeding practices to ensuring proper housing and health care. Simultaneously, ensuring animal welfare involves addressing the physical and psychological needs of animals, promoting humane treatment, and adhering to ethical standards. As the global demand for animal products continues to rise, the challenge for livestock producers is to balance productivity with animal welfare considerations.

## Livestock Management Practices

### 1. Nutrition Management

#### Importance of Balanced Diets

- **Nutrient Requirements:** Livestock require balanced diets that meet their specific nutritional needs for growth, reproduction, and maintenance. Essential nutrients include proteins, carbohydrates, fats, vitamins, and minerals. Understanding the nutrient requirements of different species and growth stages is crucial for optimal health and productivity.

- **Feed Formulation:** Effective feed formulation involves creating rations that provide the necessary nutrients in the right proportions. This can be achieved through the use of commercial feed additives, supplements, and by-products. Regular assessment and adjustment of feed formulations based on performance data and changing needs are essential.
- **Feeding Strategies:** Implementing feeding strategies that optimize feed efficiency and reduce waste can improve productivity and reduce costs. Techniques such as precision feeding, which adjusts feed delivery based on individual animal needs, can enhance performance and reduce environmental impact.

### 2. Housing and Environment

### Design and Management

- **Housing Systems:** Proper housing systems protect livestock from harsh weather conditions, provide comfort, and facilitate management practices. Housing systems can vary widely, from traditional barns and pens to modern, climate-controlled facilities. The design should consider factors such as ventilation, space allowance, and cleanliness.
- **Environmental Enrichment:** Providing environmental enrichment, such as bedding, toys, and space to move, supports the physical and psychological well-being of animals. Enrichment helps reduce stress and prevent behavioral issues, contributing to better overall health.
- **Waste Management:** Effective waste management practices, including manure handling and disposal, are essential for maintaining a clean environment and minimizing the risk of disease. Systems such as composting and anaerobic digestion can also provide valuable resources for fertilization and energy production.

### 3. Breeding and Genetics

### Improving Livestock Quality

- **Selective Breeding:** Selective breeding involves choosing animals with desirable traits to improve herd performance. Traits such as growth rate, reproductive efficiency, and disease resistance are targeted through breeding programs to enhance overall productivity and resilience.
- **Genetic Tools:** Advances in genetic tools, including genomic selection and marker-assisted breeding, allow for more precise and rapid improvements in livestock genetics. These tools help identify and select

animals with superior traits and accelerate the development of new breeds and varieties.

- **Genetic Diversity:** Maintaining genetic diversity within livestock populations is important for long-term health and productivity. Genetic diversity helps populations adapt to changing environmental conditions and reduces the risk of inherited diseases.

### 4. Health Management

### Preventive and Therapeutic Measures

- **Disease Prevention:** Implementing preventive health measures, including vaccinations, biosecurity protocols, and regular health screenings, helps protect livestock from diseases and reduces the need for therapeutic interventions. Biosecurity measures include controlling access to animal areas, disinfecting equipment, and managing wildlife interactions.
- **Veterinary Care:** Regular veterinary care, including routine check-ups and emergency treatment, is essential for maintaining livestock health. Veterinarians play a critical role in diagnosing and treating diseases, managing reproductive health, and providing guidance on herd management practices.
- **Monitoring and Record-Keeping:** Monitoring animal health through regular observation and maintaining detailed health records helps identify trends, track treatment outcomes, and make informed management decisions. Electronic record-keeping systems can streamline data collection and analysis.

## Livestock Welfare Practices

### 1. Humane Treatment

### Ensuring Animal Well-Being

- **Housing Conditions:** Providing adequate space, proper ventilation, and appropriate bedding contributes to the comfort and well-being of livestock. Ensuring that housing conditions meet or exceed welfare standards is crucial for preventing stress and health issues.
- **Handling Practices:** Humane handling practices involve minimizing stress and discomfort during interactions with livestock. This includes using gentle handling techniques, avoiding rough or aggressive behavior, and providing adequate training for handlers.
- **Social Interaction:** Livestock are social animals and benefit from interactions with other animals. Providing opportunities for socialization,

such as group housing or communal areas, supports mental well-being and reduces the risk of behavioral problems.

**2. Ethical Considerations**

**Addressing Welfare Concerns**

- **Animal Welfare Standards:** Adhering to established animal welfare standards and guidelines, such as those set by organizations like the World Organisation for Animal Health (OIE) and the American Veterinary Medical Association (AVMA), ensures that livestock are treated ethically and humanely.
- **Pain Management:** Implementing pain management practices, including the use of anesthetics and analgesics during procedures, helps minimize suffering and improve overall welfare. Pain assessment and management should be integrated into routine veterinary care.
- **End-of-Life Care:** Providing humane end-of-life care, including proper euthanasia procedures, is an important aspect of livestock welfare. Ensuring that euthanasia is performed humanely and with minimal stress is crucial for ethical treatment.

**3. Environmental and Social Impacts**

**Balancing Productivity with Sustainability**

- **Environmental Impact:** Livestock farming can have significant environmental impacts, including greenhouse gas emissions, land use, and water consumption. Implementing sustainable practices, such as improving feed efficiency, reducing waste, and adopting climate-smart technologies, can help mitigate these impacts.
- **Social and Economic Factors:** Livestock farming plays a vital role in rural economies and food security. Addressing social and economic factors, including fair labor practices, community engagement, and support for smallholder farmers, contributes to the overall sustainability and equity of the livestock sector.

## Innovations and Future Directions

**1. Technological Advancements**

**Enhancing Efficiency and Welfare**

- **Precision Livestock Farming:** Precision livestock farming uses technologies such as sensors, monitoring systems, and data analytics to track individual animal performance and health. This approach allows for targeted interventions and improved management practices, enhancing both productivity and welfare.

- **Automated Systems:** Automation in livestock management, including feeding systems, milking robots, and waste handling, can improve efficiency and reduce labor requirements. Automated systems can also contribute to more consistent and precise management practices.
- **Genomic and Biotechnological Innovations:** Advances in genomics and biotechnology offer new opportunities for improving livestock genetics and health. Techniques such as gene editing and genomic selection can accelerate genetic improvements and address specific challenges in livestock production.

### 2. Sustainable Practices

### Promoting Long-Term Viability

- **Circular Economy Approaches:** Adopting circular economy approaches in livestock farming, such as recycling waste into feed or fertilizer and reducing resource inputs, can enhance sustainability and reduce environmental impact.
- **Climate-Smart Practices:** Implementing climate-smart practices, including optimizing feed use, improving manure management, and adopting low-emission technologies, can help mitigate the effects of climate change and promote environmental stewardship.
- **Ethical and Transparent Supply Chains:** Enhancing transparency and ethical practices in supply chains, including certification programs and traceability systems, can improve consumer confidence and support sustainable and humane livestock production.

## 3. Policy and Regulation

Policy and regulation play a critical role in shaping livestock management and welfare practices. They establish standards and guidelines that ensure the humane treatment of animals, promote environmental sustainability, and support the efficient and ethical operation of livestock systems. This section explores the key aspects of policy and regulation in livestock management and welfare, including the development and implementation of regulations, the role of governmental and non-governmental organizations, and the impact on industry practices and animal well-being.

### 1. Regulatory Frameworks

### Development and Implementation

- **Legislative Framework:** The legislative framework for livestock management and welfare typically includes national and regional laws that set minimum standards for animal care, housing, nutrition, and health. These laws are often based on scientific research and ethical considerations and aim to protect animal welfare and ensure public safety.

- **Regulatory Agencies:** Government agencies are responsible for enforcing livestock regulations and ensuring compliance. Agencies may include departments of agriculture, veterinary services, and environmental protection. These agencies develop and implement regulations, conduct inspections, and handle violations.
- **Standards and Guidelines:** Regulatory frameworks often include detailed standards and guidelines for specific aspects of livestock management, such as housing conditions, feed quality, and veterinary care. These standards are designed to ensure that animals are treated humanely and that their basic needs are met.

### Challenges and Considerations

- **Enforcement:** Effective enforcement of regulations is crucial for ensuring compliance and protecting animal welfare. Challenges in enforcement may include limited resources, insufficient training for inspectors, and difficulties in monitoring large or remote farms.
- **Adaptability:** Regulations must be adaptable to address emerging issues and advances in technology. For example, new practices in precision livestock farming may require updates to existing regulations to ensure they address the latest developments.
- **Balancing Interests:** Policymakers must balance the interests of various stakeholders, including farmers, animal welfare advocates, and consumers. Regulations should be designed to address animal welfare concerns while also considering the economic and operational impacts on livestock producers.

## 2. Role of Governmental and Non-Governmental Organizations

### Governmental Organizations

- **National Agencies:** National agencies play a central role in developing and enforcing livestock regulations. They may issue guidelines, conduct inspections, and provide support to farmers for compliance. Examples include the U.S. Department of Agriculture (USDA) and the European Food Safety Authority (EFSA).
- **Local Authorities:** Local authorities may have specific regulations and oversight responsibilities that address regional issues and conditions. They often work in conjunction with national agencies to ensure that

regulations are applied consistently across different areas.

### Non-Governmental Organizations (NGOs)

- **Animal Welfare Organizations:** NGOs, such as the World Animal Protection and the Humane Society International, advocate for improved animal welfare standards and work to raise public awareness about livestock issues. They may also provide resources and support for implementing best practices in animal care.
- **Industry Associations:** Industry associations, such as the National Cattlemen's Beef Association or the Poultry Industry Council, often develop voluntary codes of conduct and best practice guidelines for their members. These guidelines may complement or exceed regulatory requirements and promote high standards of animal welfare.
- **Research Institutions:** Research institutions and universities conduct studies that inform policy and regulation. Their research provides evidence-based insights into animal welfare, management practices, and the impacts of various interventions.

## 3. Impact on Industry Practices and Animal Well-Being

### Positive Impacts

- **Improved Welfare Standards:** Effective regulations and policies contribute to higher welfare standards by setting minimum requirements for housing, nutrition, and health care. This leads to better living conditions for livestock and improved overall health and productivity.
- **Enhanced Consumer Confidence:** Transparent and well-enforced regulations help build consumer confidence in the safety and ethical standards of animal products. Consumers are more likely to support and purchase products from farms that adhere to recognized welfare standards.
- **Promotion of Best Practices:** Regulations and guidelines often promote the adoption of best practices in livestock management. This includes implementing humane handling procedures, providing environmental enrichment, and maintaining proper health care.

### Challenges and Areas for Improvement

- **Compliance Costs:** Compliance with regulations can involve significant costs for farmers, including investments in infrastructure, equipment, and training. Support mechanisms, such as subsidies or technical assistance, may be needed to help farmers meet regulatory requirements.

- **Regulatory Burden:** The complexity and volume of regulations can create a burden for livestock producers, especially smallholder farmers. Simplifying and streamlining regulatory processes can help reduce administrative challenges and improve compliance.
- **Global Variation:** Regulatory standards can vary significantly between countries and regions, leading to challenges in international trade and market access. Harmonizing standards and fostering international cooperation can help address these challenges and promote consistency.

## 4. Future Directions

### Innovative Approaches

- **Data-Driven Regulation:** Leveraging data and technology can enhance regulatory frameworks by providing real-time monitoring and evidence-based insights. Data-driven approaches can improve enforcement, identify emerging issues, and support more effective regulation.
- **Stakeholder Engagement:** Engaging stakeholders, including farmers, animal welfare advocates, and consumers, in the regulatory process can help ensure that regulations are practical, effective, and aligned with societal values. Collaborative approaches can foster consensus and support the development of balanced policies.
- **Global Standards:** Promoting the development of global standards for animal welfare can help address inconsistencies and facilitate international trade. Collaborative efforts between governments, NGOs, and industry organizations can drive progress towards more uniform and effective regulations.

### Policy Development and Advocacy

- **Evidence-Based Policies:** Continued investment in research and evidence-based policy development is essential for addressing emerging challenges and improving livestock welfare. Policymakers should rely on scientific research and stakeholder input to inform regulatory decisions.
- **Education and Training:** Providing education and training for farmers, regulators, and industry professionals can enhance understanding and implementation of regulations. Training programs can help ensure that all stakeholders are equipped to meet regulatory requirements and adopt best practices.
- **Continuous Improvement:** Regulatory frameworks should be regularly reviewed and updated to reflect advances in technology, changes in industry practices, and evolving societal values. Continuous

improvement ensures that regulations remain relevant and effective in promoting animal welfare.

## Conclusion

Livestock management and welfare are essential components of modern animal agriculture, impacting productivity, health, and ethical considerations. Effective management practices, including nutrition, housing, breeding, and health care, contribute to optimal performance and well-being of livestock. Ensuring humane treatment and addressing ethical concerns are critical for maintaining high standards of animal welfare.

Future directions in livestock management and welfare involve leveraging technological advancements, adopting sustainable practices, and supporting effective policy and regulation. By addressing challenges and exploring innovative solutions, the livestock sector can achieve a balance between productivity and ethical considerations, ensuring a sustainable and humane approach to animal agriculture.

Through continued research, collaboration, and commitment to best practices, the future of livestock management and welfare holds the potential for a more resilient, efficient, and compassionate agricultural system.

## References

Berckmans, D. (2014). Precision livestock farming technologies for welfare management in intensive livestock systems. Rev. Sci. Tech, 33(1), 189-196.

Coleman, G. J., & Hemsworth, P. H. (2014). Training to improve stockperson beliefs and behaviour towards livestock enhances welfare and productivity. Rev. Sci. Tech, 33, 131-137.

Orihuela, A. (2021). Management of livestock behavior to improve welfare and production. Animal, 15, 100290.

Phillips, C. J. C., Wojciechowska, J., Meng, J., & Cross, N. (2009). Perceptions of the importance of different welfare issues in livestock production. Animal, 3(8), 1152-1166.

# 6

# Agroecology and Biodiversity

**Abstract**

*Agroecology and biodiversity are interconnected concepts that play a crucial role in shaping sustainable agricultural systems. Agroecology, an approach that integrates ecological principles into agricultural practices, aims to enhance the sustainability, resilience, and productivity of farming systems. Biodiversity, encompassing the variety of life forms in ecosystems, provides essential ecosystem services and supports the stability of agroecosystems. This chapter explores the principles of agroecology, the importance of biodiversity in agriculture, and the ways in which they interact to promote sustainable farming. It also examines the challenges and future directions for integrating agroecology and biodiversity in modern agriculture, highlighting case studies and innovative practices that contribute to the advancement of sustainable agricultural systems.*

***Keywords:*** *Agroecology, biodiversity, sustainable agriculture, ecosystem services, ecological principles, agroecosystems, resilience, environmental sustainability, case studies, innovative practices.*

## Introduction

Agroecology and biodiversity are central to the development of sustainable agricultural systems that address the challenges of food security, environmental degradation, and climate change. Agroecology applies ecological concepts and principles to farming practices, aiming to create agricultural systems that are more resilient, productive, and harmonious with natural ecosystems. Biodiversity, which refers to the variety and variability of life forms within an ecosystem, is a critical component of agroecological systems, providing essential services that support crop production and ecosystem health. It examines the benefits of integrating ecological approaches into farming practices and the role of biodiversity in enhancing agricultural resilience and productivity. Additionally, it addresses the challenges and future directions for advancing agroecology and biodiversity in agriculture.

# Agroecology: Principles and Practices

## 1. Ecological Principles in Agroecology

### Integration of Ecosystem Functions

- **Ecological Relationships:** Agroecology emphasizes the understanding and management of ecological relationships between crops, soils, and other components of the agroecosystem. This includes recognizing the role of natural processes such as nutrient cycling, pest regulation, and pollination in supporting agricultural productivity.
- **Diversity and Complexity:** Agroecological systems aim to increase biodiversity and complexity within agricultural landscapes. By incorporating diverse plant species, crop varieties, and natural habitats, agroecology promotes ecosystem stability and resilience.
- **Resource Efficiency:** Efficient use of resources, including water, nutrients, and energy, is a key principle of agroecology. Techniques such as intercropping, agroforestry, and conservation tillage are employed to optimize resource use and reduce environmental impact.

### Holistic Management

Holistic management is an integrative approach to resource management that emphasizes the interconnectedness of ecological, economic, and social factors. This approach is grounded in systems thinking, which views ecosystems and agricultural systems as complex networks of interacting components. It recognizes that changes in one part of the system can have far-reaching effects throughout the entire system. Holistic management aims to optimize these interactions to achieve long-term sustainability, resilience, and productivity.

At its core, holistic management integrates several key principles. Systems thinking is fundamental, focusing on the interdependence of various system components and the importance of feedback loops that influence system behavior. This approach promotes a long-term perspective, emphasizing sustainability and adaptive management. Adaptive management involves continuously adjusting practices based on new information and changing conditions, which enhances system resilience and flexibility. Additionally, holistic management prioritizes integrated decision-making, considering ecological, economic, and social factors to ensure balanced and sustainable outcomes. Engaging stakeholders is also crucial, as it helps to incorporate diverse perspectives and build consensus around management strategies.

In practical applications, holistic management can be seen in various contexts. In agriculture, it involves practices such as rotational grazing and crop diversification, which enhance soil health, increase productivity, and support ecological balance. For land restoration, holistic management focuses on rehabilitating degraded ecosystems through coordinated efforts like reforestation and erosion control, often involving local communities to align restoration efforts with local needs and values. In water management, a holistic approach considers entire watersheds and integrates land use management, water allocation, and pollution control to optimize water resources and protect water quality.

The benefits of holistic management are significant. It enhances resilience by improving system stability and adaptive capacity, allowing systems to better withstand and recover from disturbances. It also improves resource efficiency by optimizing the use of resources such as soil and water, leading to cost savings and reduced environmental impact. Additionally, holistic management can increase productivity by integrating practices that support soil health and biodiversity, ultimately enhancing yields and farm performance.

However, implementing holistic management faces several challenges. These include the need for specialized knowledge and skills, resource constraints, and the complexity of monitoring and evaluating outcomes. Effective stakeholder engagement and conflict resolution are also necessary to address diverse priorities and ensure successful implementation. Future directions for holistic management include leveraging technological advancements, developing supportive policy frameworks, and fostering collaborative approaches. By addressing these challenges and continuing to innovate, holistic management can contribute to more sustainable and resilient systems that benefit both people and the environment.

## 2. Agroecological Practices

### Diverse Crop Systems

- **Crop Rotation and Diversity:** Implementing diverse crop rotations and polycultures helps to maintain soil fertility, reduce pest and disease pressures, and enhance overall system resilience. Diverse cropping systems can also improve yields and reduce the need for chemical inputs.
- **Agroforestry:** Agroforestry integrates trees and shrubs into agricultural landscapes, providing multiple benefits such as improved soil fertility, erosion control, and habitat for wildlife. This practice supports biodiversity and enhances ecosystem services.
- **Cover Crops:** Using cover crops during fallow periods helps to prevent soil erosion, improve soil structure, and enhance nutrient cycling. Cover crops also provide habitat for beneficial organisms and contribute to overall system health.

## Soil and Water Management

- **Conservation Tillage:** Conservation tillage practices, such as no-till and reduced tillage, help to maintain soil structure, reduce erosion, and increase water infiltration. These practices support soil health and reduce the impact of agricultural activities on the environment.
- **Integrated Water Management:** Agroecology promotes the use of integrated water management practices, including rainwater harvesting, efficient irrigation systems, and soil moisture conservation techniques. These practices help to optimize water use and reduce environmental impact.

## Pest and Disease Management

- **Biological Control:** Agroecological pest management focuses on using natural predators, parasitoids, and pathogens to control pest populations. Biological control methods reduce the reliance on chemical pesticides and support ecological balance.
- **Cultural Practices:** Implementing cultural practices such as crop rotation, plant diversity, and habitat management helps to reduce the incidence of pests and diseases. These practices contribute to a more balanced and resilient agroecosystem.

## Biodiversity in Agriculture

### 1. Importance of Biodiversity

### Ecosystem Services

- **Pollination:** Biodiversity supports pollinator populations, which are essential for the reproduction of many crop species. Healthy pollinator communities enhance crop yields and quality, contributing to food security.
- **Nutrient Cycling:** Diverse plant and microbial communities contribute to nutrient cycling and soil fertility. Biodiversity helps to maintain soil structure, enhance organic matter decomposition, and improve nutrient availability for crops.
- **Pest and Disease Regulation:** Biodiversity helps to regulate pest and disease populations by providing habitat for natural enemies and promoting ecological interactions that reduce pest pressures. Diverse agroecosystems are less susceptible to pest outbreaks and disease spread.

### Resilience and Adaptation

- **Environmental Stress Tolerance:** Biodiverse agricultural systems are more resilient to environmental stresses such as drought, floods, and temperature fluctuations. Genetic diversity within crops and livestock enhances their ability to adapt to changing conditions.
- **Ecosystem Stability:** High levels of biodiversity contribute to the stability and functioning of agroecosystems. Diverse systems are better able to withstand disturbances, recover from setbacks, and maintain productivity over time.

## 2. Promoting Biodiversity in Agriculture

### Habitat Conservation

- **Protected Areas:** Establishing protected areas within agricultural landscapes, such as riparian buffers, wetlands, and natural reserves, helps to conserve habitats and support biodiversity. These areas provide refuge for wildlife and contribute to ecosystem health.
- **Wildlife-Friendly Practices:** Implementing wildlife-friendly practices, such as maintaining hedgerows, planting cover crops, and creating wildlife corridors, supports biodiversity and enhances ecological interactions within farming systems.

### Seed and Genetic Diversity

- **Seed Saving and Exchange:** Promoting seed saving and exchange programs helps to preserve traditional crop varieties and maintain genetic diversity. Diverse seed banks and community seed networks contribute to agricultural resilience and innovation.
- **Breeding for Diversity:** Incorporating genetic diversity into breeding programs helps to develop crop varieties with enhanced traits, such as disease resistance and environmental tolerance. This approach supports long-term sustainability and adaptation.

### Agroecological Certification

- **Certification Programs:** Agroecological certification programs, such as organic certification and biodiversity-friendly labels, provide incentives for farmers to adopt practices that support biodiversity and sustainability. Certification programs help to promote and recognize environmentally friendly practices.

## Challenges and Future Directions

### 1. Challenges

#### Economic Pressures

- **Cost and Profitability:** Implementing agroecological practices and promoting biodiversity may involve higher initial costs and require changes in farming operations. Economic pressures and market demands can pose challenges for farmers adopting these approaches.
- **Market Access:** Limited access to markets for agroecologically produced goods and products can affect the financial viability of adopting biodiversity-friendly practices. Developing market opportunities and consumer awareness is essential for supporting sustainable agriculture.

#### Knowledge and Training

- **Lack of Knowledge:** Limited knowledge and understanding of agroecological principles and biodiversity conservation can hinder the adoption of sustainable practices. Training and education programs are needed to support farmers and stakeholders in implementing these approaches.
- **Research Gaps:** Research on the effectiveness and impacts of agroecological practices and biodiversity conservation is still evolving. Addressing research gaps and expanding knowledge is crucial for advancing sustainable agricultural practices.

### 2. Future Directions

#### Innovative Practices

- **Technology Integration:** Integrating technology, such as remote sensing, precision agriculture, and data analytics, can enhance the implementation and monitoring of agroecological practices and biodiversity conservation. Technology can support decision-making and improve efficiency in sustainable agriculture.
- **Participatory Approaches:** Engaging farmers, communities, and stakeholders in the development and implementation of agroecological practices promotes collaboration and ensures that practices are adapted to local conditions and needs. Participatory approaches can enhance the effectiveness and adoption of sustainable practices.

#### Policy Support

- **Incentives and Support:** Developing policies that provide incentives and support for agroecological practices and biodiversity conservation is essential for promoting sustainability. Policy measures, such as subsidies, grants, and technical assistance, can help offset costs and encourage adoption.

- **Integrated Policy Frameworks:** Creating integrated policy frameworks that address both agricultural productivity and environmental sustainability can support the transition to more sustainable farming systems. Coordinated efforts across sectors and levels of government are needed to achieve comprehensive and effective policies.

## Global Collaboration

- **International Initiatives:** Participating in international initiatives and agreements related to biodiversity and sustainable agriculture, such as the Convention on Biological Diversity (CBD) and the Sustainable Development Goals (SDGs), can drive global progress and support the integration of agroecology and biodiversity in agriculture.
- **Knowledge Sharing:** Promoting knowledge sharing and collaboration between countries, research institutions, and organizations can advance understanding and implementation of agroecological practices. Global networks and partnerships can facilitate the exchange of best practices and innovations.

## Conclusion

Agroecology and biodiversity are integral to the development of sustainable agricultural systems that address the challenges of food security, environmental degradation, and climate change. Agroecology applies ecological principles to farming practices, enhancing sustainability and resilience, while biodiversity provides essential ecosystem services and supports agroecosystem stability.

Integrating agroecology and biodiversity in agriculture offers numerous benefits, including improved productivity, enhanced resilience, and reduced environmental impact. However, challenges such as economic pressures, knowledge gaps, and limited market access must be addressed to advance these approaches.

Future directions involve adopting innovative practices, supporting policy measures, and fostering global collaboration to promote sustainable agriculture. By leveraging technology, engaging stakeholders, and developing supportive policies, the agricultural sector can achieve a more sustainable and resilient future, balancing productivity with environmental and ecological considerations.

## References

Bohan, D. A., Richter, A., Bane, M., Therond, O., & Pocock, M. J. (2022). Farmer-led agroecology for biodiversity with climate change. Trends in Ecology & Evolution, 37(11), 927-930.

Chappell, M.J., & LaValle, L. A. (2011). Food security and biodiversity: can we have both? An agroecological analysis. Agriculture and human values, 28, 3-26.

Lemanceau, P., Maron, P.A., Mazurier, S., Mougel, C., Pivato, B., Plassart, P., and Wipf, D. (2015). Understanding and managing soil biodiversity: a major challenge in agroecology. Agronomy for Sustainable Development, 35, 67-81.

Liere, H., Jha, S. and Philpott, S.M. (2017). Intersection between biodiversity conservation, agroecology, and ecosystem services. Agroecology and Sustainable Food Systems, 41(7), 723-760.

# 7

# Climate-Resilient Agriculture

**Abstract**

*Climate-resilient agriculture is an approach designed to adapt agricultural practices to the challenges posed by climate change while maintaining productivity and sustainability. This chapter explores the principles, strategies, and technologies involved in developing climate-resilient agricultural systems. It covers various aspects, including soil management, water conservation, crop diversification, and innovative technologies. The chapter also examines the role of policy, research, and community engagement in fostering climate resilience. Through a comprehensive analysis of current practices and future directions, it aims to provide a thorough understanding of how agriculture can adapt to a changing climate while ensuring food security and environmental sustainability.*

***Keywords:*** *Climate resilience, agriculture, soil management, water conservation, crop diversification, climate adaptation, sustainable agriculture, innovative technologies, policy, community engagement.*

## Introduction

Climate change presents significant challenges to agriculture, affecting crop yields, water availability, soil health, and pest dynamics. As global temperatures rise and weather patterns become more erratic, it is imperative for agricultural systems to adapt in order to ensure food security and environmental sustainability. Climate-resilient agriculture is an approach that addresses these challenges by integrating adaptive strategies and technologies to enhance the resilience of agricultural systems. This chapter provides an in-depth examination of climate-resilient agriculture, exploring its principles, strategies, and future directions.

## Principles of Climate-Resilient Agriculture

### 1. Adaptability

### Flexible Practices

- **Dynamic Adjustments:** Climate-resilient agriculture involves practices that can be adjusted based on changing climate conditions. This flexibility is crucial for responding to variability in temperature, precipitation, and other environmental factors.

- **Local Adaptation:** Adapting practices to local conditions is essential. What works in one region may not be effective in another due to differences in climate, soil, and other factors.

### Risk Management

Risk management in agriculture is a comprehensive approach to identifying, assessing, and mitigating the various risks that can impact farming operations. Given the inherent uncertainties in agriculture, including climate variability, market fluctuations, and pest outbreaks, effective risk management is crucial for maintaining both productivity and resilience.

The first step in risk management is identifying the types of risks that farmers face. Environmental risks include climate-related challenges such as extreme weather events, droughts, and floods, which can directly affect crop yields and soil health. Soil risks, such as degradation and erosion, also pose significant threats to agricultural productivity. Water risks, including variability in water availability and quality, are another critical concern, as they influence irrigation practices and overall water resource management. Biological risks, such as pests, diseases, and weeds, can cause substantial losses in crop and livestock production, while economic risks, including market price fluctuations and rising input costs, impact farmers' financial stability. Social risks, such as labor availability and policy changes, further complicate the agricultural landscape.

Once risks are identified, they must be assessed to understand their potential impact and likelihood. This involves evaluating the probability of each risk occurring and its potential effects on agricultural operations. Risk mapping, which uses spatial data and geographic information systems (GIS), helps in visualizing and analyzing vulnerabilities within agricultural systems, allowing for targeted interventions.

Mitigating these risks involves implementing various strategies tailored to each type of risk. For environmental risks, practices such as climate adaptation, soil conservation, and efficient water management are crucial. Adapting agricultural practices to changing climate conditions, using soil conservation techniques like reduced tillage and cover cropping, and optimizing water use through technologies like drip irrigation are effective strategies. Biological risks can be managed through integrated pest management (IPM), disease-resistant crop varieties, and effective weed control methods. Economic risks can be mitigated by diversifying crops and livestock, developing robust financial plans, and implementing market strategies to stabilize income. Social risks require managing labor availability and staying informed about policy changes.

Effective risk management also involves continuous monitoring and evaluation. Early warning systems, regular inspections, and real-time data analysis help in anticipating and preparing for potential risks. Evaluating the performance of risk management strategies ensures they are effective and identifies areas for improvement.

Looking ahead, technological innovations such as advanced analytics, remote sensing, and machine learning will play a significant role in enhancing risk management practices. Supportive policies and capacity building will further support farmers in managing risks effectively. Collaborative approaches, including partnerships between governments, research institutions, and farmers, will drive progress in developing and implementing effective risk management strategies. Overall, a proactive and adaptive approach to risk management is essential for building resilient agricultural systems that can withstand the challenges of a changing environment and ensure long-term sustainability.

## 2. Sustainability

### Resource Efficiency

- **Optimized Use of Resources:** Climate-resilient agriculture emphasizes the efficient use of resources such as water, soil, and nutrients. Techniques such as precision agriculture and integrated nutrient management contribute to resource efficiency.
- **Minimizing Waste:** Reducing waste and optimizing resource use help to minimize environmental impact and improve overall sustainability.

### Ecosystem Health

- **Soil Health:** Maintaining and improving soil health is fundamental to climate resilience. Practices such as cover cropping, reduced tillage, and organic matter addition enhance soil structure, fertility, and water-holding capacity.
- **Biodiversity:** Promoting biodiversity within agricultural systems supports ecosystem functions and services, such as pollination and pest control, which are vital for resilience.

## 3. Innovation

### Technological Advancements

- **Precision Agriculture:** Technologies such as GPS, drones, and sensors enable precise monitoring and management of agricultural practices, optimizing resource use and enhancing resilience.

- **Climate-Resilient Crop Varieties:** Breeding and genetically modifying crops to tolerate extreme weather conditions, pests, and diseases contribute to resilience. Improved seed varieties can enhance yield stability and reduce vulnerability.

### Knowledge and Research

- **Ongoing Research:** Continuous research into climate change impacts and adaptation strategies is essential for developing effective solutions. Collaboration between researchers, farmers, and policymakers drives innovation.
- **Knowledge Dissemination:** Sharing knowledge and best practices with farmers through extension services, training programs, and workshops supports the adoption of climate-resilient practices.

## Strategies for Climate-Resilient Agriculture

### 1. Soil Management

### Conservation Tillage

- **Reduced Tillage:** Conservation tillage practices, such as no-till or reduced tillage, help to maintain soil structure, reduce erosion, and enhance water infiltration. These practices improve soil health and resilience.
- **Organic Matter Addition:** Adding organic matter, such as compost or manure, to soil improves its fertility, water-holding capacity, and structure, contributing to resilience.

### Soil Health Monitoring

- **Regular Testing:** Monitoring soil health through regular testing helps to assess nutrient levels, pH, and other factors that influence soil quality and crop performance.
- **Soil Amendments:** Applying appropriate soil amendments based on test results ensures that soil conditions are optimized for crop growth and resilience.

### 2. Water Management

### Efficient Irrigation

- **Drip Irrigation:** Drip irrigation systems deliver water directly to the plant roots, minimizing water wastage and improving efficiency. This method is particularly useful in arid and semi-arid regions.

- **Rainwater Harvesting:** Collecting and storing rainwater for use during dry periods enhances water availability and reduces dependence on external sources.

### Water Conservation Practices

- **Mulching:** Applying mulch to the soil surface helps to retain moisture, reduce evaporation, and suppress weeds, contributing to water conservation.
- **Contour Farming:** Implementing contour farming techniques, such as building terraces and planting along contour lines, helps to reduce water runoff and soil erosion.

## 3. Crop Diversification

### Variety Selection

- **Diverse Crop Varieties:** Growing a variety of crop species and varieties helps to spread risk and improve resilience. Diverse cropping systems are less susceptible to pests, diseases, and climate extremes.
- **Climate-Resilient Crops:** Selecting crop varieties that are specifically bred for resilience to drought, heat, and other climate stresses enhances overall system stability.

### Agroforestry

- **Integrating Trees:** Agroforestry involves integrating trees and shrubs into agricultural systems, providing benefits such as improved soil fertility, reduced erosion, and enhanced biodiversity.
- **Multi-Use Systems:** Designing multi-use systems that combine crops, trees, and livestock maximizes resource use and supports resilience.

## 4. Livestock Management

### Grazing Strategies

- **Rotational Grazing:** Implementing rotational grazing systems helps to manage pasture health, reduce overgrazing, and improve soil fertility. This practice supports resilience by maintaining vegetation cover and reducing erosion.
- **Supplemental Feeding:** Providing supplemental feed and water during periods of stress ensures that livestock remain healthy and productive under challenging conditions.

### Animal Breeding

- **Resilient Breeds:** Breeding livestock for traits such as heat tolerance, disease resistance, and feed efficiency enhances their ability to withstand climate-related challenges.
- **Health Management:** Regular health checks and preventive measures help to maintain livestock well-being and productivity in the face of climate stressors.

## Role of Policy and Support

### 1. Policy Frameworks

### Supportive Policies

- **Incentives for Adoption:** Policies that provide financial incentives, subsidies, or grants for adopting climate-resilient practices encourage farmers to invest in sustainable technologies and approaches.
- **Research Funding:** Supporting research into climate adaptation and resilience through government funding and grants advances knowledge and innovation in agricultural practices.

### Climate Adaptation Plans

- **National Strategies:** Developing and implementing national climate adaptation strategies that include agricultural sectors helps to align efforts and resources for resilience building.
- **Local Adaptation:** Tailoring adaptation plans to local conditions and needs ensures that strategies are relevant and effective in specific regions.

### 2. Research and Extension

### Collaborative Research

- **Partnerships:** Collaboration between research institutions, universities, and agricultural organizations drives innovation and knowledge development in climate-resilient agriculture.
- **Applied Research:** Conducting applied research that focuses on practical solutions and strategies helps to bridge the gap between research and implementation.

### Extension Services

- **Knowledge Transfer:** Extension services play a crucial role in disseminating information and best practices to farmers. Providing training, workshops, and resources supports the adoption of climate-resilient practices.

- **Farmer Support:** Offering technical support and guidance helps farmers to implement and manage climate-resilient practices effectively.

## Challenges and Future Directions

### 1. Implementation Barriers

#### Economic Constraints

- **Cost of Technologies:** The initial cost of adopting climate-resilient technologies and practices can be a barrier for farmers, particularly in low-income regions. Financial support and incentives are needed to address this challenge.
- **Market Access:** Limited access to markets for climate-resilient products can affect the financial viability of adopting these practices. Developing market opportunities and consumer awareness is essential.

#### Knowledge Gaps

- **Lack of Information:** Limited knowledge and understanding of climate-resilient practices can hinder adoption. Expanding research and providing comprehensive training and education are crucial.
- **Data Availability:** Access to reliable data and information on climate impacts and adaptation strategies supports effective decision-making and planning.

### 2. Future Directions

#### Technological Innovations

- **Advancing Technologies:** Continued development and integration of new technologies, such as remote sensing, artificial intelligence, and climate modeling, enhance the ability to monitor, manage, and adapt agricultural systems.
- **Smart Farming:** Implementing smart farming techniques that combine data analytics and automation improves efficiency, productivity, and resilience.

#### Policy and Support Systems

- **Strengthening Policies:** Developing and implementing comprehensive policies that support climate resilience, including incentives, research funding, and adaptation plans, is essential for promoting sustainable agriculture.
- **Enhancing Collaboration:** Fostering collaboration between governments, research institutions, NGOs, and farmers helps to drive progress and address challenges in climate-resilient agriculture.

## Global and Local Integration

Global and local integration in agriculture is a strategy that harmonizes international insights with local practices to address agricultural challenges effectively and sustainably. This approach recognizes that while global perspectives provide valuable standards, innovations, and market dynamics, local knowledge is crucial for tailoring these insights to specific contexts and conditions.

Internationally, global standards and agreements, such as those established by the Food and Agriculture Organization (FAO) and climate accords like the Paris Agreement, set benchmarks for food safety, quality, and environmental sustainability. These standards guide countries in adopting practices that align with global expectations and contribute to international goals. Collaborative global research initiatives, facilitated by organizations like CGIAR, advance agricultural innovation by developing climate-resilient crops and sustainable technologies. The transfer of such innovations across borders enhances agricultural practices worldwide. Additionally, understanding global market trends and trade agreements helps align local production with international demand, while efficient global supply chains ensure effective distribution and resource management.

At the local level, integrating global knowledge with indigenous practices ensures that agricultural strategies are adapted to specific environmental, social, and economic conditions. Local knowledge, including traditional farming techniques and local climate adaptations, provides valuable insights into managing local resources and addressing unique challenges. Engaging with local communities and stakeholders ensures that agricultural practices are culturally appropriate and supported by those directly affected. Combining global and local perspectives enables the development of resilient agricultural systems that are both innovative and contextually relevant, ultimately enhancing sustainability and productivity across diverse regions.

## Conclusion

Climate-resilient agriculture is essential for adapting to the impacts of climate change while ensuring food security and environmental sustainability. By embracing principles of adaptability, sustainability, and innovation, and implementing strategies such as soil management, water conservation, and crop diversification, agricultural systems can enhance their resilience to climate-related challenges.

Addressing barriers to implementation, advancing technologies, and developing supportive policies are crucial for promoting climate-resilient agriculture. Through continued research, collaboration, and targeted support, the agricultural sector can effectively navigate the challenges of a changing climate and build systems that are both productive and sustainable.

## References

Munawar, S., ul Qamar, M.T., Mustafa, G., Khan, M. S. and Joyia, F. A. (2020). Role of biotechnology in climate resilient agriculture. Environment, climate, plant and vegetation growth, 339-365.

Rai, R. K., Bhatta, L. D., Acharya, U. and Bhatta, A. P. (2018). Assessing climate-resilient agriculture for smallholders. Environmental Development, 27, 26-33.

Reddy, P.P. (2015). Climate resilient agriculture for ensuring food security (Vol. 373). New Delhi: Springer India.

Wilson, M. L., and VanBuren, R. (2022). Leveraging millets for developing climate resilient agriculture. Current Opinion in Biotechnology, 75, 102683.

# 8

# Urban and Peri-Urban Agriculture

**Abstract**

*Urban and peri-urban agriculture (UPA) has emerged as a vital component of contemporary food systems, addressing food security, environmental sustainability, and socio-economic development in densely populated areas. This chapter explores the multifaceted dimensions of UPA, including its benefits, challenges, and future directions. It examines the roles of urban and peri-urban farming in enhancing food security, promoting environmental sustainability, and supporting local economies. Key strategies for successful implementation, including community involvement, policy support, and technological innovations, are discussed. The chapter provides a comprehensive analysis of UPA practices across various global contexts and offers recommendations for scaling up and integrating UPA into broader urban planning and development frameworks.*

***Keywords:*** *Urban agriculture, peri-urban agriculture, food security, environmental sustainability, urban planning, community involvement, policy support, technological innovations, local economies.*

## Introduction

Urban and peri-urban agriculture (UPA) refers to the cultivation of crops and the raising of animals within and around urban areas. As cities expand and populations grow, UPA has become increasingly important for enhancing food security, supporting local economies, and promoting environmental sustainability. This chapter provides an in-depth exploration of UPA, examining its significance, benefits, challenges, and future prospects. By analyzing various practices and strategies, it aims to offer insights into how UPA can be effectively integrated into urban planning and development.

## Historical Context and Evolution of UPA

### Early Practices

Urban and peri-urban agriculture (UPA) has a rich historical background that reflects its deep roots in human civilization. The practice of growing food within and around urban areas is not a modern phenomenon but has evolved over centuries, adapting to changing socio-economic conditions and technological advancements. Understanding the early practices of UPA provides valuable insights into its current role and future potential.

## 1. Ancient Civilizations

The origins of urban agriculture can be traced back to ancient civilizations where agricultural practices were integrated into city life. Early urban societies recognized the importance of food production within their settlements to support growing populations and ensure food security.

- **Mesopotamia:** In ancient Mesopotamia (circa 3000 BCE), early urban centers like Uruk and Babylon utilized irrigation systems to cultivate crops in and around their cities. The development of canal systems allowed for efficient water management and supported agriculture in the arid environment.
- **Ancient Egypt:** The ancient Egyptians practiced agriculture along the Nile River, using its annual flooding to enrich the soil and support crop production. Urban areas such as Thebes and Memphis had surrounding agricultural lands that provided essential food resources.
- **The Roman Empire:** Roman cities such as Rome and Pompeii had extensive agricultural activities within their city limits. Gardens and small farms supplied fresh produce to urban populations. The Romans also implemented advanced agricultural techniques, such as crop rotation and the use of manure as fertilizer.

## 2. Medieval and Renaissance Europe

During the medieval period, urban and peri-urban agriculture continued to play a significant role in European cities, adapting to the feudal system and the growth of trade.

- **Medieval Cities:** In medieval Europe, cities like London and Paris incorporated gardens and small farms within their city walls. Monasteries and abbeys often had extensive gardens and orchards that provided food for their inhabitants and for trade.
- **Renaissance Innovations:** The Renaissance period saw advancements in agricultural techniques and urban planning. The use of enclosed gardens, or "potagers," became common in cities. These gardens were used for growing vegetables, herbs, and medicinal plants, contributing to urban food supply.

## 3. Colonial and Early Modern Periods

As European powers established colonies and expanded their territories, the practice of urban agriculture was introduced to new regions, adapting to different climates and cultures.

- **Colonial America:** In colonial America, urban agriculture was vital for early settlers. Cities like New York and Philadelphia had community gardens and small farms that provided fresh produce and livestock to urban residents.
- **Industrial Revolution:** The Industrial Revolution brought rapid urbanization, leading to the growth of cities and a shift away from traditional farming. However, urban agriculture persisted in various forms, such as allotment gardens and city farms, as a means to supplement food supplies and provide recreational space.

### 4. 20th Century Developments

The 20th century witnessed significant changes in urban and peri-urban agriculture, driven by industrialization, technological advancements, and changing social dynamics.

- **Victory Gardens:** During World War I and World War II, "victory gardens" were promoted in urban areas to encourage citizens to grow their own food as part of the war effort. These gardens were established in vacant lots, parks, and private yards, contributing to urban food security and boosting morale.
- **Urban Farming Movements:** The latter half of the 20th century saw the rise of urban farming movements, particularly in response to growing concerns about food security, environmental sustainability, and community health. Initiatives such as community gardens, rooftop farms, and guerrilla gardening became popular in cities around the world.

### 5. Early Practices in Non-Western Regions

Urban agriculture has a long history in non-Western regions as well, where traditional practices have shaped contemporary approaches.

- **Asia:** In countries like China and Japan, urban agriculture has been practiced for centuries. In ancient China, cities had gardens and small farms that provided vegetables and herbs. In Japan, "shinden" gardens were integrated into urban and rural landscapes, contributing to local food systems.
- **Latin America:** Indigenous cultures in Latin America practiced urban agriculture through techniques such as terracing and raised beds. In cities like Mexico City, ancient agricultural practices were integrated into urban environments, including the use of cinemas (man-made farming islets) for crop production.

## Modern Developments

In the 20th century, urban agriculture experienced a decline due to industrialization and urbanization. However, the late 20th and early 21st centuries have seen a resurgence of interest in UPA, driven by concerns about food security, environmental sustainability, and the desire for local food systems. Innovations in technology and changes in urban policies have facilitated the growth of urban and peri-urban farms, integrating them into modern cityscapes.

## Benefits of Urban and Peri-Urban Agriculture

### 1. Enhancing Food Security

### Local Food Production

- **Fresh Produce Availability:** UPA provides access to fresh fruits, vegetables, and herbs, reducing reliance on long-distance food transportation and ensuring that urban populations have access to nutritious food.
- **Food Resilience:** By diversifying food sources and reducing dependence on external supply chains, UPA enhances food resilience and security in urban areas, particularly during crises or supply disruptions.

### Economic Impact

Urban and peri-urban agriculture (UPA) has a profound economic impact on cities and surrounding areas, influencing various aspects of local economies, food systems, and community well-being. By integrating agriculture into urban and peri-urban environments, these practices create economic opportunities, enhance food security, and contribute to sustainable development. The following sections explore the multifaceted economic impacts of UPA.

### 1. Job Creation and Economic Opportunities

- **Employment Generation:** UPA creates diverse employment opportunities across various sectors, including farming, processing, distribution, and retail. Jobs are generated in areas such as urban farms, community gardens, farmer's markets, and agribusinesses. This employment contributes to reducing urban unemployment and providing livelihood options for residents.
- **Entrepreneurship:** UPA encourages entrepreneurial activities by providing opportunities for small-scale businesses. Urban farmers, food producers, and agribusinesses can develop niche markets for locally grown produce, value-added products, and agritourism experiences. This fosters local economic development and innovation.

- **Skills Development:** Engaging in UPA helps individuals acquire new skills related to agriculture, horticulture, and business management. Training and educational programs associated with UPA initiatives equip participants with practical skills that can enhance their employability and entrepreneurial prospects.

**2. Food Security and Cost Savings**

- **Local Food Production:** By producing food locally, UPA reduces reliance on long supply chains and minimizes transportation costs. This local food production helps stabilize food prices and provides fresh, affordable produce to urban populations. It also contributes to food security by diversifying food sources and reducing vulnerability to external market fluctuations.
- **Cost Savings for Consumers:** UPA can lead to cost savings for consumers by reducing the costs associated with transporting and storing food. Locally grown produce is often less expensive than imported alternatives, benefiting households with lower food costs.
- **Community Resilience:** UPA enhances community resilience by providing a buffer against food supply disruptions. In times of economic downturns or crises, urban agriculture can offer a reliable source of food and reduce dependence on external food sources.

**3. Real Estate and Property Value**

- **Property Value Enhancement:** The presence of urban farms, community gardens, and green spaces can positively impact property values in surrounding areas. Green and productive landscapes contribute to the aesthetic appeal and desirability of neighborhoods, potentially increasing property values and attracting investment.
- **Urban Regeneration:** UPA projects can be instrumental in urban regeneration efforts, transforming vacant or underutilized land into productive agricultural spaces. This revitalization enhances the attractiveness of urban areas and encourages further investment and development.
- **Reduced Infrastructure Costs:** Integrating agriculture into urban planning can reduce infrastructure costs related to waste management and storm water management. Green spaces and urban farms contribute to environmental sustainability and can lower the need for costly infrastructure improvements.

## 4. Local Economic Development

- **Stimulating Local Economies:** UPA stimulates local economies by creating demand for inputs, services, and products within the community. Local suppliers, including nurseries, tool manufacturers, and packaging companies, benefit from the increased demand generated by urban agriculture activities.
- **Boosting Local Markets:** Urban farms and community gardens often sell their produce at local farmers' markets, grocery stores, and restaurants. This supports local markets and strengthens the local food economy, benefiting both producers and consumers.
- **Tourism and Recreation:** UPA projects can attract tourists and visitors interested in exploring urban farms, community gardens, and agritourism experiences. This tourism generates additional revenue for local businesses and contributes to the local economy.

## 5. Environmental and Health Cost Savings

- **Environmental Benefits:** UPA practices contribute to environmental sustainability by promoting green spaces, reducing transportation-related emissions, and improving urban biodiversity. These environmental benefits can lead to cost savings related to pollution control, climate adaptation, and urban heat island effects.
- **Health Cost Savings:** Access to fresh, nutritious produce from urban agriculture can improve public health and reduce healthcare costs associated with diet-related diseases. Healthier populations contribute to lower healthcare expenses and improved quality of life.

## 6. Challenges and Considerations

- **Economic Viability:** The economic viability of UPA projects can be influenced by factors such as land costs, regulatory challenges, and market competition. Ensuring the financial sustainability of UPA initiatives requires careful planning, investment, and support.
- **Market Access:** Urban farmers may face challenges accessing markets and competing with large-scale agricultural producers. Developing effective marketing strategies and distribution channels is essential for the success of UPA projects.
- **Funding and Support:** Securing funding and support for UPA initiatives can be challenging. Public and private investments, grants, and policy support are crucial for the development and sustainability of urban agriculture projects.

## 2. Promoting Environmental Sustainability

### Green Spaces and Urban Cooling

- **Green Infrastructure:** Integrating UPA into urban environments creates green spaces that contribute to the aesthetic and environmental quality of cities. These spaces can help reduce the urban heat island effect and improve overall air quality.
- **Waste Management:** UPA systems often incorporate waste recycling practices, such as composting organic waste, which reduces landfill use and improves soil fertility.

### Resource Efficiency

- **Water Management:** Urban farms use innovative water management techniques, including rainwater harvesting and efficient irrigation systems, to minimize water use and promote sustainability.
- **Energy Use:** UPA practices often focus on energy efficiency, such as using renewable energy sources and optimizing energy use in vertical farms and greenhouses.

## 3. Supporting Local Economies and Community Engagement

### Community Development

- **Social Cohesion:** UPA fosters community engagement and social cohesion by bringing people together to work on shared projects. Community gardens and urban farms become focal points for social interaction and collaboration.
- **Educational Opportunities:** UPA provides educational opportunities for urban residents, including children and adults, to learn about food production, environmental stewardship, and sustainable practices.

### Local Food Systems

- **Short Supply Chains:** By producing food locally, UPA reduces the distance food travels from farm to table, supporting local food systems and strengthening connections between producers and consumers.
- **Economic Resilience:** Local food systems supported by UPA can help cities become more economically resilient by diversifying food sources and supporting local businesses.

## Challenges of Urban and Peri-Urban Agriculture

### 1. Land Use and Space Constraints

### Limited Space

- **Urban Density:** High population density and limited available land in urban areas present challenges for establishing and expanding UPA projects. Creative solutions, such as vertical farming and rooftop gardens, are needed to maximize space usage.
- **Land Competition:** Urban land is often in demand for commercial and residential development, making it difficult for UPA initiatives to secure and maintain space.

### Zoning Regulations

- Land Use Policies: Zoning regulations and land use policies may restrict agricultural activities in urban areas. Advocating for policy changes and incorporating UPA into urban planning frameworks are essential for overcoming these challenges.

### 2. Environmental and Health Concerns

### Pollution

- **Soil and Air Quality:** Urban areas may have higher levels of soil and air pollution, which can impact the safety and quality of produce grown in UPA systems. Monitoring and managing environmental contaminants are critical for ensuring food safety.
- **Water Contamination:** Water used in UPA systems may be at risk of contamination from urban runoff or industrial activities. Implementing water filtration and monitoring systems helps mitigate these risks.

### Pest and Disease Management

- **Pest Infestations:** Urban environments may harbor pests and diseases that can affect UPA crops. Integrated pest management practices and regular monitoring are necessary to manage and control these issues.

### 3. Economic and Operational Challenges

Urban and peri-urban agriculture (UPA) offers significant benefits, including enhanced food security, local economic development, and environmental sustainability. However, it also faces a range of economic and operational challenges that can impact its effectiveness and sustainability. Addressing these challenges requires a nuanced understanding of the issues and proactive strategies for overcoming them.

## 1. Economic Challenges

- **High Land Costs:** One of the most significant economic challenges for UPA is the high cost of urban land. In densely populated cities, land is often expensive and in short supply, making it difficult for urban farmers to secure affordable plots for agriculture. This challenge is exacerbated in cities where real estate development and commercial interests dominate.
- **Access to Capital:** Funding is crucial for establishing and maintaining UPA projects. However, many urban agriculture initiatives struggle to secure sufficient capital for initial setup, infrastructure, and ongoing operations. Access to financial resources may be limited, especially for small-scale or community-based projects.
- **Economic Viability:** The economic viability of UPA can be uncertain, particularly when it comes to generating sufficient revenue to cover operational costs and achieve profitability. Factors such as market competition, fluctuating produce prices, and the cost of inputs (seeds, fertilizers, etc.) can impact the financial stability of UPA projects.
- **Market Access and Distribution:** Urban farmers may face challenges accessing markets and establishing distribution channels for their products. Competing with established agricultural suppliers and navigating urban supply chains can be difficult, especially for smaller producers.
- **Regulatory and Policy Barriers:** In some cities, regulatory and policy barriers can impede the growth of UPA. Zoning laws, land use regulations, and health and safety standards may not always accommodate agricultural activities within urban areas, creating obstacles for urban farmers.

## 2. Operational Challenges

- **Limited Space:** Space constraints are a major operational challenge in urban environments. Urban farms and community gardens often operate in limited areas, which can restrict the scale and scope of agricultural activities. Effective space management and innovative farming techniques, such as vertical farming and rooftop gardens, are essential for maximizing productivity.
- **Resource Management:** Managing resources efficiently is critical for successful UPA. Urban farmers must contend with issues such as limited water supply, soil quality, and waste management. Implementing sustainable practices, such as rainwater harvesting and composting, can help address these challenges but may require additional effort and investment.

- **Soil Quality and Contamination:** Urban soils may be contaminated with pollutants or have poor quality, posing challenges for growing healthy crops. Testing and remediation of contaminated soils can be costly and time-consuming. In some cases, urban farms may use raised beds or container gardening to mitigate soil quality issues.
- **Pest and Disease Management:** Urban environments can present unique challenges for pest and disease management. High population densities and proximity to various urban activities may increase the risk of pest infestations and disease outbreaks. Integrated pest management strategies and regular monitoring are essential for addressing these issues.
- **Infrastructure and Logistics:** Efficient infrastructure and logistics are crucial for UPA operations. Urban farms need access to reliable transportation, storage facilities, and utilities. Managing these logistical aspects within a city's infrastructure can be complex and require careful planning.
- **Climate and Weather:** Urban agriculture must contend with local microclimates and weather variations, which can impact crop growth and yield. For example, urban heat islands can lead to higher temperatures, affecting plant health. Adapting to local climatic conditions and utilizing weather data for planning are essential for managing these challenges.

**3. Social and Community Challenges**

- **Community Engagement:** Building and maintaining community support is vital for the success of UPA projects. Engaging local residents, fostering participation, and addressing community needs and preferences are crucial for creating successful urban agriculture initiatives.
- **Equity and Accessibility:** Ensuring equitable access to UPA opportunities and resources is important for promoting social inclusion. Addressing disparities in access to land, capital, and training can help ensure that UPA benefits are shared broadly within urban communities.
- **Education and Training:** Providing education and training for urban farmers and community members is essential for effective UPA. Access to knowledge about farming techniques, business management, and sustainable practices can enhance the success and impact of urban agriculture projects.

**4. Technological and Innovation Challenges**

- **Technology Integration:** Integrating new technologies into UPA can be challenging, particularly for small-scale or community-based projects.

The adoption of technologies such as automated systems, sensors, and data analytics requires investment and technical expertise.

- **Research and Development:** Ongoing research and development are necessary to address the specific needs and challenges of urban agriculture. Investing in R&D can lead to innovations in farming techniques, crop varieties, and resource management practices.
- **Scalability:** Scaling up successful UPA initiatives to reach larger audiences or multiple locations can be complex. Developing scalable models that maintain efficiency and effectiveness is a key challenge for expanding the impact of urban agriculture.

### 5. Policy and Regulatory Challenges

- **Zoning and Land Use Regulations:** Zoning laws and land use regulations may not always accommodate agricultural activities in urban areas. Advocating for policy changes and developing flexible zoning solutions can help overcome these barriers.
- **Health and Safety Standards:** Ensuring compliance with health and safety standards is crucial for UPA projects, especially those involving food production. Navigating regulatory requirements and maintaining high standards for food safety and quality are essential for protecting public health.
- **Supportive Policies:** Developing and implementing supportive policies that promote UPA can help address economic and operational challenges. Public policies that provide financial incentives, technical support, and regulatory flexibility can enhance the viability and impact of urban agriculture.

## Strategies for Successful Urban and Peri-Urban Agriculture

### 1. Policy and Planning

### Supportive Policies

- **Urban Agriculture Integration:** Incorporating UPA into urban planning and development policies helps to ensure that agricultural activities are supported and integrated into city infrastructure.
- **Incentives and Grants:** Providing incentives, grants, and subsidies for UPA projects can encourage investment and support the growth of urban and peri-urban farming initiatives.

### Zoning and Land Use

Zoning and land use regulations are critical in urban and peri-urban agriculture (UPA), influencing where and how agricultural activities can occur within cities. Restrictive zoning laws may limit the availability of suitable land for farming, while land use policies may not always accommodate agricultural practices. This can hinder the establishment and expansion of UPA projects. Advocating for flexible zoning regulations and integrating agriculture into urban planning are essential to overcome these barriers. Adjusting policies to support diverse land uses, including urban farms and community gardens, can enhance the feasibility and success of UPA initiatives.

## 2. Community Engagement and Education

### Community Involvement

- **Participatory Planning:** Involving community members in the planning and decision-making processes for UPA projects ensures that initiatives meet local needs and preferences.
- **Volunteer Programs:** Engaging volunteers in UPA activities fosters community ownership and support, and provides opportunities for skill development and education.

### Educational Programs

- **Workshops and Training:** Offering workshops and training programs on urban farming techniques, sustainable practices, and business management helps build the capacity of urban farmers.
- **School and Youth Programs:** Incorporating UPA into school curricula and youth programs promotes awareness and interest in agriculture and sustainability among younger generations.

## 3. Technological Innovations

### Advanced Farming Techniques

- **Vertical Farming:** Vertical farming systems utilize stacked layers to grow crops in urban environments, optimizing space and resource use. These systems can be implemented in buildings, warehouses, and other structures.
- **Hydroponics and Aquaponics:** Hydroponic and aquaponic systems use nutrient-rich water to grow plants without soil, offering efficient and sustainable alternatives for urban agriculture.

### Data and Analytics

- **Precision Agriculture:** Leveraging data and analytics to monitor and optimize crop growth, resource use, and environmental conditions enhances the efficiency and productivity of UPA systems.

- **Smart Technologies:** Implementing smart technologies, such as sensors, automation, and remote monitoring, improves the management and operation of urban farms.

## Future Directions

### 1. Scaling Up UPA

#### Replication and Expansion

- **Successful Models:** Identifying and replicating successful UPA models from different cities can provide valuable insights and strategies for scaling up and expanding urban agriculture initiatives.
- **Collaborative Networks:** Building networks and partnerships among urban farmers, policymakers, and researchers facilitates knowledge sharing and collaboration for scaling up UPA projects.

### 2. Integration with Urban Planning

#### Holistic Approaches

- **Urban-Rural Linkages:** Strengthening connections between urban and peri-urban areas through integrated planning and resource management supports the sustainability of UPA systems.
- **Green Infrastructure:** Incorporating UPA into green infrastructure plans enhances the environmental and social benefits of urban agriculture, contributing to overall urban sustainability.

### 3. Research and Innovation

Research and innovation are pivotal in advancing urban and peri-urban agriculture (UPA), addressing the challenges of food security, environmental sustainability, and resource efficiency. Through targeted research and the development of innovative technologies and practices, UPA can evolve to meet the needs of growing urban populations while contributing to broader sustainability goals. This section explores key areas of research and innovation that are shaping the future of UPA.

#### Advances in Agricultural Technologies

#### Vertical Farming

- **Technology Overview:** Vertical farming involves growing crops in vertically stacked layers or towers, often within controlled indoor environments. This technology optimizes space usage, allowing for high-density food production in urban areas.
- **Innovations:** Advances in vertical farming include the development of automated systems for planting, harvesting, and climate control.

Integration of artificial intelligence (AI) and machine learning enables precise control of environmental conditions and resource use, improving crop yields and reducing operational costs.

- **Benefits:** Vertical farming reduces the land footprint required for agriculture, minimizes transportation needs, and enhances food security by localizing production. It also allows for year-round crop cultivation and efficient use of water and nutrients.

### Hydroponics and Aquaponics

- **Hydroponics:** Hydroponic systems grow plants in nutrient-rich water rather than soil. Innovations in hydroponic technology include the use of advanced nutrient delivery systems, energy-efficient lighting, and automated monitoring.
- **Aquaponics:** Aquaponics combines hydroponics with aquaculture, where fish waste provides nutrients for plants, and plants help to filter and purify water for fish. Recent advancements focus on optimizing system integration, improving water quality management, and enhancing the efficiency of nutrient cycles.
- **Benefits:** Both hydroponics and aquaponics offer efficient water and nutrient use, reduced soil degradation, and the ability to grow crops in urban environments with limited space. These systems are also scalable and adaptable to various urban settings.

### Smart Agriculture Technologies

- **Sensors and IoT:** The Internet of Things (IoT) and sensor technologies provide real-time data on environmental conditions, soil moisture, and crop health. Smart sensors help monitor and manage various factors, such as temperature, humidity, and nutrient levels.
- **Automation:** Automated systems, including robotic planters, harvesters, and irrigation controllers, enhance the efficiency and precision of UPA operations. Automation reduces labor requirements and minimizes human error.
- **Data Analytics:** Advanced data analytics and AI algorithms analyze data collected from sensors to optimize growing conditions, predict crop yields, and manage resources. Predictive models support decision-making and enhance productivity.

## Sustainable Practices and Resource Management

### Waste Management and Recycling

- **Composting:** Research on composting technologies focuses on improving the efficiency of organic waste decomposition and enhancing compost quality. Innovations include the development of composting systems that integrate with urban waste management infrastructure.
- **Waste-to-Resource Technologies:** Technologies that convert organic waste into valuable resources, such as biogas or biochar, contribute to sustainable waste management and resource recovery. These technologies support circular economy principles in urban agriculture.

### Water Conservation and Management

- **Efficient Irrigation Systems:** Innovations in irrigation technology, such as drip irrigation and misting systems, improve water use efficiency and reduce water waste. Advanced systems include automated irrigation controllers that adjust water delivery based on real-time data.
- **Rainwater Harvesting:** Research on rainwater harvesting technologies focuses on optimizing collection, storage, and filtration systems. Rainwater harvesting systems can supplement water supply for urban farms and reduce reliance on municipal water sources.

## Social and Economic Innovations

### Community Engagement and Education

- **Participatory Research:** Engaging communities in research and development processes ensures that UPA technologies and practices are relevant and meet local needs. Participatory research approaches involve community members in designing, testing, and implementing UPA projects.
- **Educational Programs:** Innovations in educational programs, such as interactive workshops and online courses, increase awareness and knowledge of UPA practices. These programs target diverse audiences, including school children, urban residents, and aspiring farmers.

### Business Models and Economic Strategies

- **Local Food Networks:** Research on local food networks explores models for connecting urban farmers with consumers, such as community-supported agriculture (CSA) and farm-to-table initiatives. These models support local economies and enhance food accessibility.
- **Economic Viability:** Innovations in business models for UPA include subscription services, direct-to-consumer sales, and value-

added products. These models help urban farmers achieve financial sustainability and expand their market reach.

## Policy and Regulatory Research

### Policy Development

- **Supportive Regulations:** Research on policy development focuses on creating supportive regulatory frameworks for UPA. This includes zoning regulations, land use policies, and incentives that promote urban agriculture and integrate it into urban planning.
- **Public-Private Partnerships:** Innovations in policy research explore partnerships between public institutions, private companies, and community organizations to support UPA initiatives. Collaborative approaches leverage resources, expertise, and funding for successful implementation.

### Impact Assessment

- **Environmental Impact:** Research on the environmental impact of UPA assesses the effects of urban farming practices on soil health, water quality, and biodiversity. This research informs best practices and policies for minimizing negative environmental impacts.
- **Social Impact:** Evaluating the social impact of UPA includes assessing its effects on community cohesion, public health, and access to nutritious food. Impact assessments help demonstrate the value of UPA and support its integration into urban development plans.

## Conclusion

Urban and peri-urban agriculture plays a crucial role in enhancing food security, promoting environmental sustainability, and supporting local economies in rapidly growing cities. By integrating global perspectives with local knowledge, and addressing the challenges and opportunities of UPA, cities can build more resilient and sustainable food systems. Effective policies, community engagement, and technological innovations are key to advancing UPA and ensuring its successful integration into urban planning and development frameworks. Through continued research, collaboration, and adaptation, UPA can contribute to creating more sustainable and equitable urban environments.

## References

Appeaning Addo, K. (2010). Urban and peri-urban agriculture in developing countries studied using remote sensing and in situ methods. Remote Sensing, 2(2), 497-513.

Graefe, S., Buerkert, A., & Schlecht, E. (2019). Trends and gaps in scholarly literature on urban and peri-urban agriculture. Nutrient Cycling in Agroecosystems, 115, 143-158.

Petts, J. (2012). The economics of urban and peri-urban agriculture. In Continuous productive urban landscapes (pp. 65-77). Routledge.

Thebo, A. L., Drechsel, P., & Lambin, E. F. (2014). Global assessment of urban and peri-urban agriculture: irrigated and rainfed croplands. Environmental Research Letters, 9(11), 114002.

# 9

# Digital Transformation in Agriculture

**Abstract**

*Digital transformation in agriculture represents a profound shift towards integrating digital technologies into farming practices to enhance productivity, sustainability, and efficiency. This chapter explores the multifaceted aspects of digital transformation in agriculture, including the role of technologies such as the Internet of Things (IoT), artificial intelligence (AI), big data analytics, and block chain. It examines how these technologies are reshaping agricultural practices, improving decision-making, and addressing challenges related to resource management, crop monitoring, and supply chain optimization. The chapter also discusses the implications of digital transformation for stakeholders, including farmers, policymakers, and technology providers, and offers recommendations for leveraging digital tools to advance sustainable agricultural practices and ensure food security in the digital age.*

***Keywords:*** *Digital transformation, agriculture, Internet of Things (IoT), artificial intelligence (AI), big data analytics, block chain, precision farming, smart agriculture, supply chain optimization, resource management.*

## Introduction

Digital transformation in agriculture is revolutionizing the way farming is practiced by integrating advanced technologies to optimize productivity, enhance sustainability, and improve efficiency. As the global population continues to grow and the demand for food increases, the agriculture sector faces mounting pressures to produce more with fewer resources. Digital technologies offer innovative solutions to these challenges, enabling farmers to make data-driven decisions, monitor crop health in real-time, and streamline supply chains. This chapter delves into the various aspects of digital transformation in agriculture, exploring its impact on farming practices, the technologies driving this change, and the future prospects for a digitally enabled agricultural sector.

## Key Technologies Driving Digital Transformation

### 1. Internet of Things (IoT)

#### 1.1 Technology Overview

- **Definition and Components:** The Internet of Things (IoT) refers to the network of interconnected devices that collect and exchange data through sensors and communication technologies. In agriculture, IoT devices include sensors, drones, and smart equipment that monitor environmental conditions, crop health, and soil moisture.
- **Applications:** IoT technology is used for precision farming, where sensors collect data on soil conditions, weather patterns, and crop growth. This data is analyzed to optimize irrigation, fertilization, and pest control, leading to more efficient and sustainable farming practices.

#### 1.2 Benefits

- **Real-Time Monitoring:** IoT enables real-time monitoring of agricultural conditions, allowing farmers to make timely decisions and respond to issues such as pest infestations or nutrient deficiencies.
- **Resource Efficiency:** By providing precise data on resource use, IoT helps reduce waste and optimize the application of water, fertilizers, and pesticides, promoting sustainable agriculture.

#### 1.3 Challenges

- **Data Security:** The increased connectivity of IoT devices raises concerns about data security and privacy. Ensuring robust cybersecurity measures is essential to protect sensitive agricultural data.
- **Infrastructure:** Implementing IoT solutions requires reliable internet connectivity and infrastructure, which may be limited in rural or remote areas.

### 2. Artificial Intelligence (AI)

#### 2.1 Technology Overview

- **Definition and Components:** Artificial intelligence (AI) involves the use of algorithms and machine learning to analyze data, make predictions, and automate processes. In agriculture, AI is applied to various tasks, including crop forecasting, pest detection, and autonomous machinery.
- **Applications:** AI algorithms process data from sensors, drones, and satellite imagery to predict crop yields, identify plant diseases, and optimize farming practices. AI-powered tools can also automate tasks such as planting, harvesting, and weeding.

### 2.2 Benefits

- **Enhanced Decision-Making:** AI provides actionable insights based on data analysis, helping farmers make informed decisions and improve crop management practices.
- **Increased Efficiency:** AI-powered automation reduces the need for manual labor, increases operational efficiency, and lowers production costs.

### 2.3 Challenges

- **Data Quality:** The effectiveness of AI relies on the quality and accuracy of the data used for training algorithms. Ensuring high-quality data is critical for accurate predictions and insights.
- **Adoption Barriers:** The adoption of AI technologies may be hindered by high initial costs, the need for specialized skills, and resistance to change from traditional farming practices.

## 3. Big Data Analytics

### 3.1 Technology Overview

- **Definition and Components:** Big data analytics involves processing and analyzing large volumes of data to uncover patterns, trends, and insights. In agriculture, big data encompasses data from IoT devices, satellite imagery, weather forecasts, and market trends.
- **Applications:** Big data analytics is used to optimize crop management, forecast market demand, and analyze supply chain performance. By integrating diverse data sources, farmers can gain a comprehensive understanding of factors affecting crop production and market dynamics.

### 3.2 Benefits

- **Improved Forecasting:** Big data analytics enables accurate forecasting of crop yields, market prices, and weather conditions, helping farmers plan and manage their operations more effectively.
- **Strategic Planning:** Analyzing large datasets provides insights into long-term trends and patterns, allowing farmers and policymakers to make strategic decisions for sustainable agriculture.

### 3.3 Challenges

- **Data Integration:** Integrating data from multiple sources can be complex and requires robust data management systems and analytics platforms.

- **Skill Requirements:** Effective use of big data analytics requires specialized skills and expertise in data science and analytics, which may be limited in some agricultural regions.

## 4. Blockchain Technology

### 4.1 Technology Overview

- **Definition and Components:** Blockchain technology is a decentralized digital ledger that records transactions across a distributed network of computers. In agriculture, blockchain is used to enhance transparency, traceability, and security in the supply chain.
- **Applications:** Blockchain enables the tracking of agricultural products from farm to table, providing a transparent and immutable record of each transaction. This enhances food safety, reduces fraud, and improves supply chain efficiency.

### 4.2 Benefits

- **Traceability:** Blockchain technology improves traceability by providing a secure and transparent record of each step in the supply chain, from production to consumption.
- **Fraud Prevention:** By creating an immutable record of transactions, blockchain reduces the risk of fraud and ensures the authenticity of agricultural products.

### 4.3 Challenges

- **Implementation Costs:** Implementing block chain technology can be costly and requires significant investment in infrastructure and technology.
- **Complexity:** The complexity of block chain systems and the need for widespread adoption and integration can pose challenges for implementation in the agricultural sector.

## Impact of Digital Transformation on Agriculture

## 1. Enhanced Productivity and Efficiency

### 1.1 Precision Farming

- **Optimized Resource Use:** Digital technologies enable precision farming practices that optimize the use of resources such as water, fertilizers, and pesticides. By applying resources precisely where and when they are needed, farmers can increase productivity and reduce environmental impact.

- **Yield Improvement:** Real-time data and analytics help farmers monitor crop health and make data-driven decisions to improve yield and quality. Precision farming techniques also enable more accurate planting and harvesting, further enhancing productivity.

### 1.2 Automation

- **Labor Reduction:** Automation technologies, including autonomous tractors, drones, and robotic systems, reduce the need for manual labor and increase operational efficiency. Automated systems can perform tasks such as planting, weeding, and harvesting with greater precision and speed.
- **Cost Savings:** Automation helps reduce labor costs and operational expenses, making farming more cost-effective and competitive.

## 2. Improved Sustainability and Environmental Impact

### 2.1 Resource Efficiency

- **Water Conservation:** Digital technologies enable efficient water management through precision irrigation systems and real-time monitoring of soil moisture. This reduces water waste and supports sustainable water use practices.
- **Reduced Chemical Use:** By providing precise data on crop needs, digital tools help minimize the use of fertilizers and pesticides, reducing their environmental impact and promoting sustainable farming practices.

### 2.2 Environmental Monitoring

- **Climate Adaptation:** Digital technologies allow for better monitoring and adaptation to climate change by providing insights into weather patterns, soil conditions, and environmental impacts. This helps farmers adjust their practices to cope with changing climate conditions.
- **Biodiversity Conservation:** Data-driven approaches support biodiversity conservation by identifying and protecting critical habitats, monitoring wildlife, and reducing the impact of farming on natural ecosystems.

## 3. Enhanced Food Security and Supply Chain Management

### 3.1 Food Traceability

- **Transparency:** Blockchain and other digital technologies enhance food traceability, ensuring that consumers have access to accurate information about the origin and safety of their food. This builds trust and confidence in the food supply chain.

- **Recall Management:** Improved traceability facilitates faster and more efficient recall management in the event of food safety issues, reducing the risk of widespread contamination.

### 3.2 Supply Chain Optimization

- **Efficient Logistics:** Digital tools optimize supply chain logistics by providing real-time data on inventory levels, transportation routes, and market demand. This reduces inefficiencies, minimizes waste, and improves the flow of goods from farm to market.
- **Demand Forecasting:** Advanced analytics help forecast market demand and adjust production schedules accordingly, reducing surplus and shortages and improving supply chain performance.

## Challenges and Barriers to Digital Transformation

### 1. Infrastructure and Connectivity

### 1.1 Rural Connectivity

- **Internet Access:** Reliable internet connectivity is essential for implementing digital technologies in agriculture. In rural areas, limited or unstable internet access can hinder the adoption of digital tools and services.
- **Infrastructure Development:** Developing the necessary infrastructure, such as data centers and network coverage, is crucial for supporting digital transformation in agriculture.

### 2. Cost and Investment

### 2.1 Initial Investment

- **High Costs:** The initial investment required for digital technologies, including equipment, software, and training, can be prohibitive for small-scale farmers and developing regions.
- **Financing:** Access to financing and investment is essential for supporting the adoption and scaling of digital technologies in agriculture.

### 3. Skills and Training

### 3.1 Technical Expertise

- **Skill Gaps:** The effective use of digital technologies requires specialized skills and knowledge in areas such as data analysis, machine learning, and technology management. Addressing skill gaps through training and education is essential for successful digital transformation.

- **Capacity Building:** Providing capacity-building programs and resources for farmers and agricultural professionals helps bridge the skills gap and support the adoption of digital tools.

## 4. Data Privacy and Security

### 4.1 Data Protection

- **Cybersecurity:** The increased use of digital technologies raises concerns about data privacy and cybersecurity. Ensuring robust security measures and protecting sensitive agricultural data is critical for maintaining trust and protecting against cyber threats.
- **Regulation:** Developing and implementing regulations for data protection and privacy helps safeguard information and address concerns related to digital transformation.

# Future Directions

## 1. Integration and Interoperability

### 1.1 System Integration

- **Seamless Integration:** Future developments in digital agriculture will focus on integrating various technologies and systems to create a cohesive and interoperable ecosystem. This includes integrating IoT devices, AI algorithms, and blockchain systems for a comprehensive approach to agriculture.
- **Data Standardization:** Establishing data standards and protocols will facilitate the exchange and integration of data across different platforms and technologies, improving the effectiveness of digital tools.

## 2. Innovation and Research

### 2.1 Emerging Technologies

- **Advanced AI:** Research into advanced AI technologies, such as deep learning and neural networks, will drive innovation in crop monitoring, pest detection, and decision-making.
- **Next-Generation IoT:** Innovations in IoT technology, including improved sensors and communication protocols, will enhance data collection and analysis capabilities.

### 2.2 Collaborative Research

- **Public-Private Partnerships:** Collaborative research between public institutions, private companies, and research organizations will advance digital transformation in agriculture and address challenges through innovative solutions.

- **Global Collaboration:** International collaboration and knowledge sharing will support the development and adoption of digital technologies across different regions and agricultural contexts.

## 3. Policy and Regulation

### 3.1 Supportive Policies

- **Policy Frameworks:** Developing supportive policies and regulations that promote the adoption of digital technologies and address barriers will facilitate digital transformation in agriculture.
- **Incentives and Support:** Providing incentives, subsidies, and support programs for digital technology adoption will encourage investment and innovation in the agricultural sector.

## Conclusion

Digital transformation is revolutionizing agriculture by integrating advanced technologies that enhance productivity, sustainability, and efficiency. Technologies such as IoT, AI, big data analytics, and blockchain are reshaping farming practices, improving decision-making, and addressing challenges related to resource management and supply chain optimization. Despite challenges such as infrastructure limitations, high costs, and data security concerns, the potential benefits of digital transformation are substantial. By investing in research, fostering innovation, and developing supportive policies, the agricultural sector can harness the power of digital tools to create a more resilient, productive, and sustainable food system for the future.

## References

Dayıoğlu, M. A., & Turker, U. (2021). Digital transformation for sustainable future-agriculture 4.0: A review. Journal of Agricultural Sciences, 27(4), 373-399.

López-Morales, J. A., Martínez, J. A., & Skarmeta, A. F. (2020). Digital transformation of agriculture through the use of an interoperable platform. Sensors, 20(4), 1153.

Rimma, Z., Marina, T., Olga, R., Andrey, C., & Albina, S. (2020, March). Major Trends in the Digital Transformation of Agriculture. In "New Silk Road: Business Cooperation and Prospective of Economic Development"(NSRBCPED 2019) (pp. 271-275). Atlantis Press.

Shamin, A., Frolova, O., Makarychev, V., Yashkova, N., Kornilova, L., & Akimov, A. (2019, October). Digital transformation of agricultural industry. In IOP Conference Series: Earth and Environmental Science (Vol. 346, No. 1, p. 012029). IOP Publishing.

# 10

# Economic and Policy Considerations

**Abstract**

*Economic and policy considerations play a crucial role in shaping the landscape of modern agriculture. As the sector faces increasing pressures from global population growth, climate change, and resource constraints, understanding the economic dynamics and developing effective policies are essential for fostering sustainable agricultural practices. This chapter explores the economic factors influencing agriculture, including market dynamics, financial viability, and investment strategies. It also examines key policy considerations, such as regulatory frameworks, subsidies, and international trade agreements. By analyzing these aspects, the chapter aims to provide a comprehensive overview of how economic and policy considerations impact agricultural development and sustainability. It offers recommendations for policymakers, stakeholders, and industry leaders to create a more resilient and equitable agricultural sector.*

***Keywords:*** *Economic considerations, policy frameworks, agriculture, market dynamics, financial viability, investment strategies, subsidies, international trade agreements, regulatory frameworks, sustainable agriculture.*

## Introduction

The agricultural sector is fundamental to global food security and economic stability, yet it faces numerous challenges related to economic pressures and policy constraints. As the world's population continues to grow and environmental concerns mount, the need for effective economic strategies and sound policy frameworks becomes increasingly critical. This chapter delves into the economic and policy considerations that influence agricultural development, examining the interplay between economic factors, policy decisions, and agricultural outcomes. By addressing these considerations, the chapter aims to provide insights and recommendations for advancing sustainable and resilient agricultural systems.

## Economic Considerations in Agriculture

### 1. Market Dynamics

#### 1.1 Supply and Demand

- **Market Forces:** The agricultural sector is heavily influenced by supply and demand dynamics, which determine crop prices, production levels, and profitability. Changes in consumer preferences, population growth, and global trade patterns can significantly impact market conditions.
- **Price Volatility:** Agricultural markets are subject to price volatility due to factors such as weather conditions, disease outbreaks, and geopolitical events. Understanding and managing price volatility is crucial for ensuring the stability and profitability of agricultural enterprises.

#### 1.2 Global Trade

- **Trade Agreements:** International trade agreements play a significant role in shaping agricultural markets by influencing export and import dynamics, tariff rates, and market access. Trade agreements can open up new markets for agricultural products but may also introduce competition and regulatory challenges.
- **Trade Barriers:** Non-tariff barriers, such as sanitary and phytosanitary regulations, can affect agricultural trade by imposing restrictions and standards on imports and exports. Navigating these barriers is essential for accessing global markets and maintaining competitiveness.

### 2. Financial Viability and Investment

#### 2.1 Investment Strategies

- **Capital Requirements:** The agricultural sector requires significant investment in infrastructure, technology, and resources. Capital requirements can be a barrier to entry for small-scale farmers and emerging agribusinesses. Understanding investment needs and sources is essential for fostering growth and innovation.
- **Funding Sources:** Various funding sources, including private investment, government subsidies, and development programs, support agricultural development. Identifying and accessing appropriate funding sources is critical for financing projects and achieving long-term sustainability.

#### 2.2 Risk Management

- **Financial Risks:** Agricultural enterprises face financial risks related to price fluctuations, input costs, and credit availability. Effective risk management strategies, such as insurance and hedging, help mitigate financial uncertainties and protect against losses.

- **Diversification:** Diversifying agricultural activities and income sources can reduce financial risks and enhance resilience. Strategies such as crop diversification, value-added products, and alternative revenue streams contribute to financial stability and sustainability.

### 3. Economic Sustainability

#### 3.1 Resource Efficiency

- **Sustainable Practices:** Economic sustainability in agriculture is closely linked to resource efficiency. Implementing sustainable farming practices, such as precision agriculture and conservation tillage, reduces resource use and environmental impact while improving profitability.
- **Waste Reduction:** Reducing waste and optimizing resource use contribute to economic sustainability by lowering production costs and enhancing operational efficiency. Strategies include minimizing post-harvest losses and improving supply chain management.

#### 3.2 Rural Development

- **Economic Growth:** Agriculture is a key driver of rural development, contributing to job creation, income generation, and community well-being. Supporting rural infrastructure, education, and access to markets is essential for promoting economic growth and reducing poverty in rural areas.
- **Inclusive Growth:** Ensuring inclusive growth involves addressing disparities in access to resources, technology, and market opportunities. Policies and programs that support smallholder farmers, women, and marginalized communities contribute to equitable economic development.

## Policy Considerations in Agriculture

### 1. Regulatory Frameworks

#### 1.1 Environmental Regulations

- **Sustainability Standards:** Regulatory frameworks often include environmental sustainability standards that govern land use, water management, and pollution control. Compliance with these regulations ensures that agricultural practices align with environmental goals and contribute to long-term sustainability.
- **Policy Instruments:** Various policy instruments, such as environmental impact assessments, land use planning, and conservation incentives,

support the implementation of environmental regulations and promote sustainable agricultural practices.

### 1.2 Labor and Safety Regulations

- **Worker Protection:** Labor regulations ensure fair wages, safe working conditions, and protection for agricultural workers. Implementing and enforcing labor standards contribute to the well-being of workers and improve overall productivity.
- **Health and Safety:** Regulations related to health and safety, including pesticide use and food safety standards, protect consumers and workers. Compliance with these regulations is essential for maintaining public trust and ensuring the safety of agricultural products.

## 2. Subsidies and Support Programs

### 2.1 Agricultural Subsidies

- **Types of Subsidies:** Agricultural subsidies, including direct payments, price supports, and crop insurance, provide financial assistance to farmers and support specific agricultural activities. Subsidies can stabilize income, promote production, and address market failures.
- **Impact:** The impact of subsidies on agricultural markets and sustainability varies. While subsidies can provide short-term support, they may also create market distortions, encourage overproduction, and have environmental implications.

### 2.2 Support Programs

- **Research and Extension Services:** Government support programs for research and extension services enhance agricultural productivity and innovation. Funding for research, technology transfer, and extension services supports the development and dissemination of new technologies and practices.
- **Training and Education:** Support programs that provide training and education for farmers contribute to skill development, knowledge sharing, and adoption of best practices. Access to training and education improves farm management and operational efficiency.

## 3. International Trade and Agreements

### 3.1 Trade Policies

- **Export and Import Policies:** Trade policies, including tariffs, quotas, and trade agreements, shape the flow of agricultural products across borders. Policies that facilitate trade and reduce barriers support market access and competitiveness.

- **Trade Disputes:** Trade disputes and conflicts can impact agricultural markets and create uncertainties for farmers and businesses. Resolving trade disputes and negotiating fair trade agreements are crucial for maintaining stable and equitable trade relations.

### 3.2 Global Cooperation

- **International Organizations:** International organizations, such as the World Trade Organization (WTO) and the Food and Agriculture Organization (FAO), play a role in shaping global agricultural policies and fostering cooperation among countries. Participation in global initiatives and adherence to international standards contribute to sustainable agricultural development.
- **Global Challenges:** Addressing global challenges, such as climate change and food security, requires coordinated efforts and collaboration among countries. International cooperation and policy alignment are essential for addressing shared challenges and achieving common goals.

## Challenges and Opportunities

### 1. Policy Implementation

### 1.1 Policy Gaps

- **Inconsistent Policies:** Inconsistent or poorly designed policies can undermine agricultural development and sustainability. Addressing policy gaps and ensuring coherence across different policy areas is essential for effective implementation.
- **Enforcement:** Effective enforcement of policies and regulations is crucial for achieving desired outcomes. Strengthening enforcement mechanisms and addressing compliance issues contribute to successful policy implementation.

### 1.2 Policy Innovation

- **Adaptive Policies:** Developing adaptive policies that respond to changing conditions and emerging challenges is important for maintaining relevance and effectiveness. Policymakers should focus on innovative approaches and evidence-based decision-making.
- **Stakeholder Engagement:** Engaging stakeholders, including farmers, industry representatives, and community organizations, in policy development and implementation ensures that policies address real needs and challenges.

## 2. Economic Adaptation

### 2.1 Market Adaptation

- **Innovation:** Adapting to changing market conditions requires innovation and flexibility. Embracing new technologies, exploring alternative markets, and diversifying product offerings contribute to economic resilience and competitiveness.
- **Consumer Trends:** Understanding and responding to consumer trends, such as demand for organic and sustainably produced food, can create new opportunities and drive market growth.

### 2.2 Financial Resilience

- **Risk Management:** Building financial resilience through risk management strategies, such as insurance and hedging, helps mitigate financial uncertainties and protect against adverse events.
- **Investment in Technology:** Investing in technology and innovation enhances productivity and efficiency, supporting long-term economic sustainability and growth.

## Future Directions

## 1. Policy Development

### 1.1 Integrated Policies

- **Holistic Approaches:** Future policy development should focus on integrated approaches that address economic, environmental, and social dimensions of agriculture. Holistic policies that consider the interconnections between different aspects of agricultural systems can achieve more comprehensive and sustainable outcomes.
- **Evidence-Based Decision-Making:** Policymakers should prioritize evidence-based decision-making and data-driven approaches to develop effective and targeted policies. Investing in research and data collection supports informed policy development and implementation.

## 2. Economic Strategies

### 2.1 Sustainable Investment

- **Green Investments:** Promoting green investments that support sustainable agricultural practices and technologies is crucial for achieving long-term economic and environmental goals. Incentives and funding mechanisms for green investments can drive innovation and sustainability.

- **Public-Private Partnerships:** Strengthening public-private partnerships can enhance investment in agriculture and support collaborative efforts to address challenges and seize opportunities.

## 2.2 Rural Development

- **Inclusive Growth:** Fostering inclusive growth through targeted policies and programs that support smallholder farmers, rural communities, and marginalized groups contributes to equitable economic development and social well-being.
- **Infrastructure Development:** Investing in rural infrastructure, such as transportation, education, and healthcare, supports economic growth and improves the quality of life in rural areas.

## Conclusion

Economic and policy considerations are pivotal in shaping the future of agriculture. Understanding market dynamics, investment strategies, and financial viability is essential for fostering sustainable agricultural development. Effective policy frameworks, including regulatory standards, subsidies, and trade agreements, play a crucial role in supporting agricultural growth and addressing challenges. By addressing economic and policy considerations, stakeholders can promote a more resilient, equitable, and sustainable agricultural sector. Future efforts should focus on integrated policies, evidence-based decision-making, and sustainable investment strategies to ensure long-term success and resilience in agriculture.

## References

Andrew, C. O., & Alvarez, J. (1982). Adoption of agricultural technology: developments in agro-socio-economic thought. Social and economic studies, 171-189.

Kiesel, C., Dannenberg, P., Hulke, C., Kairu, J., Diez, J. R., & Sandhage-Hofmann, A. (2022). An argument for place-based policies: The importance of local agro-economic, political and environmental conditions for agricultural policies exemplified by the Zambezi region, Namibia. Environmental Science & Policy, 129, 137-149.

Milanović, M. R., & Cvijanović, D. (2009). Problems and Possibilities of Economic Evaluation of Agro-Ecological Resources. Economics of Agriculture, 56(Thematic i), 5-30.

Supriyono, J. (2018). Managing Green Agro-Industry: Economic, Environmental And Social Consideration. Faculty of Agriculture, 1.

# 11

# Education and Knowledge Transfer

**Abstract**

*Education and knowledge transfer are vital components in advancing agricultural practices and ensuring the sustainability of food systems. This chapter explores the importance of education and knowledge transfer in agriculture, focusing on how they contribute to improved farming practices, technological adoption, and sustainable development. It examines various educational approaches, including formal education, extension services, and informal learning networks, as well as the role of technology in facilitating knowledge dissemination. Additionally, the chapter discusses the challenges and opportunities associated with knowledge transfer, such as bridging the digital divide, ensuring relevance and accessibility, and fostering collaboration between stakeholders. By highlighting successful case studies and providing recommendations for enhancing education and knowledge transfer, the chapter aims to offer a comprehensive overview of how these processes can drive agricultural innovation and development.*

***Keywords:*** *Education, knowledge transfer, agriculture, extension services, technology, sustainable development, informal learning, digital divide, stakeholder collaboration, agricultural innovation.*

## Introduction

Education and knowledge transfer are fundamental to the advancement and sustainability of agriculture. As the agricultural sector faces increasing challenges such as climate change, resource constraints, and evolving market demands, the need for effective education and knowledge dissemination becomes paramount. Education equips farmers, agricultural professionals, and stakeholders with the skills and knowledge necessary to adopt innovative practices, improve productivity, and enhance sustainability. Knowledge transfer, on the other hand, ensures that valuable information and technologies reach those who can benefit from them, fostering innovation and development across the sector. This chapter explores the multifaceted aspects of education and knowledge transfer in agriculture, examining the methods, challenges, and opportunities involved.

## Educational Approaches in Agriculture

### 1. Formal Education

#### 1.1 Agricultural Colleges and Universities

- **Programs and Curricula:** Agricultural colleges and universities offer formal education programs that provide in-depth knowledge and skills in various agricultural disciplines. These programs include undergraduate and graduate degrees, as well as specialized certificates and diplomas in areas such as crop science, animal husbandry, and agricultural engineering.
- **Research and Development:** Academic institutions play a crucial role in research and development, contributing to the advancement of agricultural technologies and practices. Research conducted at these institutions often leads to innovations and improvements that benefit the broader agricultural community.

#### 1.2 Vocational Training

- **Technical Skills:** Vocational training programs focus on practical skills and hands-on experience, preparing individuals for specific roles in agriculture. These programs often include training in areas such as machinery operation, pest management, and soil conservation.
- **Industry Partnerships:** Collaboration between vocational training institutions and industry partners ensures that training programs align with current industry needs and technological advancements. This partnership helps bridge the gap between education and practical application.

### 2. Extension Services

#### 2.1 Role of Agricultural Extension

- **Knowledge Dissemination:** Agricultural extension services are designed to provide farmers and rural communities with up-to-date information and technical support. Extension agents act as intermediaries between researchers and farmers, facilitating the transfer of knowledge and technology.
- **Training and Workshops:** Extension services often organize training sessions, workshops, and field demonstrations to educate farmers on best practices, new technologies, and sustainable farming techniques. These activities help bridge the knowledge gap and promote the adoption of innovative practices.

### 2.2 Community-Based Approaches

- **Local Expertise:** Community-based extension approaches leverage local knowledge and expertise to address specific agricultural challenges. By involving community members in the development and delivery of extension services, these approaches ensure that solutions are relevant and culturally appropriate.
- **Participatory Methods:** Participatory methods, such as farmer field schools and peer learning groups, encourage active involvement and collaboration among farmers. These methods promote the exchange of knowledge and experiences, leading to practical and context-specific solutions.

## 3. Informal Learning Networks

### 3.1 Peer Learning and Networking

- **Farmer Networks:** Informal learning networks, such as farmer groups and cooperatives, provide opportunities for peer learning and knowledge exchange. Farmers share their experiences, challenges, and successes, fostering a collaborative learning environment.
- **Online Platforms:** Digital platforms and social media offer additional avenues for informal learning and knowledge sharing. Online forums, blogs, and webinars allow farmers and agricultural professionals to connect with experts and peers, access information, and discuss relevant topics.

### 3.2 Knowledge Sharing Initiatives

- **Field Days and Expos:** Field days and agricultural expos provide opportunities for farmers to learn about new technologies, products, and practices. These events facilitate direct interaction with experts, industry representatives, and other farmers.
- **Demonstration Projects:** Demonstration projects showcase innovative practices and technologies in real-world settings. By observing and participating in these projects, farmers gain practical insights and hands-on experience.

# Challenges in Knowledge Transfer

## 1. Bridging the Digital Divide

### 1.1 Access to Technology

- **Rural Connectivity:** Limited access to technology and internet connectivity in rural areas can hinder knowledge transfer and access to digital resources. Addressing connectivity issues and providing affordable technology solutions are essential for bridging the digital divide.

- **Technology Adoption:** Even when technology is available, farmers may face challenges in adopting and using it effectively. Training and support are needed to ensure that farmers can utilize digital tools and platforms to their advantage.

### 1.2 Language and Literacy Barriers

- **Language Diversity:** Language barriers can impede effective knowledge transfer, especially in regions with diverse linguistic backgrounds. Providing information in multiple languages and using culturally appropriate communication methods can improve accessibility.
- **Literacy Levels:** Low literacy levels among some farmers may limit their ability to engage with written materials and digital resources. Developing user-friendly educational materials and offering practical, hands-on training can address literacy challenges.

## 2. Ensuring Relevance and Accessibility

### 2.1 Context-Specific Solutions

- **Local Adaptation:** Knowledge and technologies must be adapted to local conditions and contexts to be effective. Tailoring educational content and practices to the specific needs and challenges of different regions ensures relevance and applicability.
- **Feedback Mechanisms:** Implementing feedback mechanisms allows farmers and stakeholders to provide input on the effectiveness and relevance of educational programs. This feedback helps refine and improve the content and delivery of knowledge transfer initiatives.

### 2.2 Inclusivity and Equity

- **Marginalized Groups:** Ensuring that educational and knowledge transfer initiatives are inclusive and equitable is crucial for reaching marginalized and underserved communities. Programs should address the needs of women, youth, and smallholder farmers to promote broad-based development.
- **Access to Resources:** Providing access to resources such as training materials, technical support, and financial assistance helps ensure that all farmers can benefit from knowledge transfer initiatives.

## Opportunities in Knowledge Transfer

### 1. Leveraging Technology

#### 1.1 Digital Tools and Platforms

- **E-Learning:** E-learning platforms and mobile applications offer flexible and accessible education options for farmers. These tools provide access to a wide range of resources, including videos, tutorials, and interactive modules.
- **Data Analytics:** Data analytics can enhance knowledge transfer by providing insights into farming practices, market trends, and environmental conditions. Analyzing data helps identify best practices and tailor recommendations to specific contexts.

#### 1.2 Innovations in Communication

- **Multimedia Resources:** Using multimedia resources, such as videos, infographics, and interactive tools, can enhance the effectiveness of knowledge transfer by making information more engaging and easier to understand.
- **Virtual Reality and Augmented Reality:** Virtual reality (VR) and augmented reality (AR) technologies offer immersive learning experiences, allowing farmers to explore and practice new techniques in a virtual environment.

### 2. Strengthening Collaboration

#### 2.1 Public-Private Partnerships

- **Collaborative Projects:** Public-private partnerships can drive innovation and improve knowledge transfer by combining resources, expertise, and networks. Collaborative projects can address specific agricultural challenges and promote the adoption of new technologies.
- **Industry Engagement:** Engaging industry stakeholders, including technology providers, agribusinesses, and research institutions, enhances the relevance and impact of knowledge transfer initiatives. Industry involvement ensures that educational programs align with current market needs and technological advancements.

#### 2.2 Regional and Global Networks

- **Knowledge Exchange:** Regional and global networks facilitate knowledge exchange and collaboration among farmers, researchers, and policymakers. Participating in these networks provides access to best practices, research findings, and successful case studies from different regions.

- **Cross-Border Collaboration:** Cross-border collaboration and knowledge sharing contribute to addressing common agricultural challenges and advancing sustainable practices. International partnerships support the transfer of innovations and solutions across borders.

## Case Studies and Best Practices

### 1. Successful Education Programs

#### 1.1 The Farmer Field School Approach

- **Program Overview:** The Farmer Field School (FFS) approach involves group-based learning where farmers participate in hands-on activities and discussions on various agricultural topics. FFS programs have been successful in improving farmers' knowledge, skills, and practices.
- **Impact:** Case studies from countries such as Kenya and Indonesia show that FFS programs have led to increased crop yields, better pest management, and enhanced farmer empowerment.

#### 1.2 Agricultural Extension Services in India

- **Program Overview:** India's agricultural extension services provide comprehensive support to farmers through training, demonstrations, and technical assistance. The services are delivered through a network of extension officers and agricultural institutions.
- **Impact:** Extension services in India have improved agricultural productivity, introduced new technologies, and supported sustainable farming practices. Successful programs include the National Rural Employment Guarantee Scheme and the Pradhan Mantri Krishi Sinchai Yojana.

### 2. Effective Knowledge Transfer Models

#### 2.1 Digital Platforms for Agricultural Education

- **Platform Overview:** Digital platforms such as e-extension services and online agricultural forums offer a range of educational resources and tools for farmers. Examples include the e-Agriculture platform and the FAO's e-learning modules.
- **Impact:** These platforms have expanded access to information, provided training on best practices, and facilitated knowledge exchange among farmers and experts.

### 2.2 Community-Based Knowledge Sharing

- **Program Overview:** Community-based knowledge sharing initiatives, such as farmer cooperatives and local agricultural networks, foster peer learning and collaboration. Programs in regions like Latin America and Sub-Saharan Africa highlight the success of community-based approaches.
- **Impact:** Community-based initiatives have enhanced local knowledge, promoted innovative practices, and strengthened social networks among farmers.

## Recommendations

### 1. Enhancing Educational Programs

### 1.1 Curriculum Development

- **Contextual Relevance:** Develop curricula that are relevant to local agricultural conditions and challenges. Incorporate practical, hands-on training and real-world applications to ensure that educational programs are effective and applicable.
- **Continuous Improvement:** Regularly update curricula and educational materials to reflect the latest research, technologies, and best practices. Engage with industry experts and stakeholders to keep programs current and relevant.

### 1.2 Expanding Access

- **Infrastructure Investment:** Invest in infrastructure to improve access to educational resources, especially in rural areas. This includes expanding internet connectivity, providing digital devices, and establishing local training centers.
- **Inclusive Programs:** Design educational programs that address the needs of diverse groups, including women, youth, and marginalized communities. Ensure that programs are accessible and inclusive, promoting equitable participation.

### 2. Improving Knowledge Transfer

### 2.1 Technology Integration

- **Digital Tools:** Leverage digital tools and platforms to facilitate knowledge transfer and reach a wider audience. Develop user-friendly applications and resources that are accessible to farmers with varying levels of digital literacy.

- **Innovative Methods:** Explore innovative methods, such as VR and AR, to enhance knowledge transfer and provide immersive learning experiences. Invest in research and development to explore new technologies and approaches.

## 2.2 Strengthening Collaboration

- **Partnerships:** Build strong partnerships between public institutions, private sector actors, and community organizations to support knowledge transfer initiatives. Collaborative efforts can enhance the effectiveness and reach of educational programs.
- **Knowledge Networks:** Establish and participate in regional and global knowledge networks to share best practices, research findings, and successful case studies. Foster cross-border collaboration to address common challenges and promote sustainable practices.

## Conclusion

Education and knowledge transfer are critical drivers of agricultural innovation and sustainability. By investing in formal education, extension services, and informal learning networks, stakeholders can enhance the skills and knowledge of farmers and agricultural professionals. Addressing challenges such as the digital divide, ensuring relevance and accessibility, and fostering collaboration are essential for effective knowledge transfer. Through successful case studies and best practices, it is clear that targeted educational programs and innovative knowledge transfer models can lead to significant improvements in agricultural practices and outcomes. By continuing to support and advance education and knowledge transfer, the agricultural sector can achieve greater productivity, sustainability, and resilience.

## References

Malhan, I. V., & Rao, S. (2007). Agricultural knowledge transfer in India: A study of prevailing communication channels. Library Philosophy and Practice, 200.

Movahedi, R., Soleimanian-Boroujeni, J., & Ketabi, M. (2021). Rural women's agricultural knowledge and information systems about agro-food processing including extension education factors. International Journal of Education Economics and Development, 12(4), 381-396.

Popescu, G. (2017). The transfer of knowledge as a source of future development. Competitiveness of Agro-Food and Environmental Economy, 6-22.

Taylor, M., & Bhasme, S. (2018). Model farmers, extension networks and the politics of agricultural knowledge transfer. Journal of Rural Studies, 64, 1-10.

# 12

# Emerging Technologies in Agriculture and the Future of Food Production

**Abstract**

*Emerging technologies are reshaping agriculture and revolutionizing food production systems. This chapter explores the latest technological advancements in agriculture, including precision farming, artificial intelligence, biotechnology, robotics, and digital platforms. It examines how these technologies are being implemented to address the challenges of modern agriculture, such as increasing productivity, enhancing sustainability, and ensuring food security. The chapter also discusses the future directions of food production, considering the impact of technological innovations on global food systems, resource management, and environmental sustainability. By analyzing current trends, opportunities, and potential challenges, the chapter provides a comprehensive overview of how emerging technologies are poised to transform agriculture and the future of food production.*

***Keywords:*** *Emerging technologies, agriculture, food production, precision farming, artificial intelligence, biotechnology, robotics, digital platforms, sustainability, food security.*

## Introduction

The agricultural sector is undergoing a profound transformation driven by emerging technologies. As the global population continues to grow and environmental challenges intensify, the need for innovative solutions to enhance food production, resource management, and sustainability becomes increasingly urgent. Emerging technologies offer new possibilities for addressing these challenges, promising to improve productivity, reduce environmental impact, and ensure food security. This chapter provides an in-depth exploration of the key emerging technologies in agriculture and their potential impact on the future of food production. It covers advancements in precision farming, artificial intelligence (AI), biotechnology, robotics, and digital platforms, and examines their implications for the agricultural industry and global food systems.

# Emerging Technologies in Agriculture

## 1. Precision Farming

### 1.1 Overview

- **Definition:** Precision farming, also known as precision agriculture, refers to the use of technology to optimize crop production and resource management. It involves the collection and analysis of data to make informed decisions about planting, fertilization, irrigation, and pest management.
- **Technologies:** Key technologies in precision farming include GPS, Geographic Information Systems (GIS), remote sensing, and variable rate technology. These tools enable farmers to monitor field conditions, apply inputs more efficiently, and reduce waste.

### 1.2 Applications

- **Soil and Crop Monitoring:** Precision farming technologies allow for detailed monitoring of soil conditions and crop health. Sensors and drones provide real-time data on soil moisture, nutrient levels, and plant growth, enabling targeted interventions.
- **Variable Rate Application:** Variable rate technology enables farmers to apply fertilizers, pesticides, and water at varying rates across different areas of a field. This approach maximizes efficiency and minimizes environmental impact.

### 1.3 Benefits and Challenges

- **Benefits:** Precision farming improves crop yields, reduces input costs, and minimizes environmental impact by optimizing resource use. It also enhances decision-making through data-driven insights.
- **Challenges:** The adoption of precision farming technologies can be limited by high costs, technical complexity, and the need for specialized skills. Additionally, data management and integration pose challenges for effective implementation.

## 2. Artificial Intelligence (AI) and Machine Learning

### 2.1 Overview

- **Definition:** AI and machine learning involve the development of algorithms and models that enable computers to learn from data and make predictions or decisions. In agriculture, AI and machine learning are used to analyze large volumes of data, identify patterns, and automate decision-making processes.

- Technologies: AI technologies in agriculture include computer vision, natural language processing, and predictive analytics. These tools are applied in various areas, such as crop monitoring, yield prediction, and disease detection.

## 2.2 Applications

- **Crop Monitoring and Management:** AI-powered image recognition and computer vision systems can identify crop diseases, pests, and nutrient deficiencies from images captured by drones or cameras. This enables timely and targeted interventions.
- **Yield Prediction:** Machine learning algorithms analyze historical and real-time data to predict crop yields and assess the impact of environmental factors. Accurate yield predictions support better planning and resource allocation.

## 2.3 Benefits and Challenges

- **Benefits:** AI and machine learning enhance accuracy, efficiency, and scalability in agricultural operations. They enable data-driven decision-making and predictive analytics, improving overall farm management.
- **Challenges:** Implementing AI technologies requires access to high-quality data, computational resources, and specialized expertise. Data privacy and security concerns also need to be addressed.

# 3. Biotechnology

## 3.1 Overview

- **Definition:** Biotechnology involves the use of biological processes and organisms to develop new agricultural products and technologies. It includes genetic modification, molecular breeding, and synthetic biology.
- **Technologies:** Key biotechnological advancements in agriculture include genetically modified (GM) crops, CRISPR gene editing, and bioengineered microbes. These technologies aim to improve crop traits, enhance resistance to pests and diseases, and increase nutritional value.

## 3.2 Applications

- **Genetically Modified Crops:** GM crops are engineered to possess specific traits, such as resistance to pests, tolerance to herbicides, or improved nutritional content. These traits contribute to higher yields and reduced reliance on chemical inputs.

- **Gene Editing:** CRISPR and other gene-editing technologies allow for precise modifications of plant genomes, enabling the development of crops with desired characteristics. This technology holds promise for accelerating breeding programs and addressing agricultural challenges.

### 3.3 Benefits and Challenges

- **Benefits:** Biotechnology offers significant potential for improving crop performance, enhancing food security, and reducing environmental impact. It also supports the development of novel agricultural products and solutions.
- **Challenges:** The adoption of biotechnology is often met with regulatory hurdles, public concerns about safety, and ethical considerations. Addressing these challenges requires transparent communication and robust regulatory frameworks.

## 4. Robotics and Automation

### 4.1 Overview

- **Definition:** Robotics and automation involve the use of robots and automated systems to perform tasks traditionally done by humans. In agriculture, these technologies are applied to tasks such as planting, harvesting, and crop monitoring.
- **Technologies:** Key advancements in agricultural robotics include autonomous tractors, robotic harvesters, and drones for aerial monitoring. These technologies improve efficiency, precision, and labor productivity.

### 4.2 Applications

- **Autonomous Vehicles:** Autonomous tractors and machinery can perform tasks such as planting, tilling, and spraying with minimal human intervention. These systems enhance efficiency and reduce labor requirements.
- **Robotic Harvesters:** Robotic harvesters are designed to pick fruits and vegetables with precision and speed. They reduce the need for manual labor and minimize crop damage during harvesting.

### 4.3 Benefits and Challenges

- **Benefits:** Robotics and automation increase productivity, reduce labor costs, and improve operational efficiency. They also enable precision farming practices and support large-scale operations.
- **Challenges:** The high initial cost of robotics and automation technology can be a barrier to adoption, particularly for small and medium-sized

farms. Additionally, integrating these technologies into existing systems requires careful planning and training.

## 5. Digital Platforms and Big Data

### 5.1 Overview

- **Definition:** Digital platforms and big data refer to the use of digital tools and large datasets to collect, analyze, and leverage information for decision-making in agriculture. These technologies enable data-driven insights and support smart farming practices.
- **Technologies:** Digital platforms include farm management software, data analytics tools, and cloud-based solutions. Big data encompasses large volumes of information from various sources, such as sensors, satellites, and market data.

### 5.2 Applications

- **Farm Management Software:** Farm management software helps farmers track and manage various aspects of their operations, including crop planning, resource allocation, and financial management. These tools support efficient and data-driven decision-making.
- **Data Analytics:** Big data analytics provides insights into trends, patterns, and correlations that inform agricultural practices. Analyzing data from sensors, weather forecasts, and market trends helps optimize operations and improve outcomes.

### 5.3 Benefits and Challenges

- **Benefits:** Digital platforms and big data enhance decision-making, improve efficiency, and support predictive analytics. They enable farmers to make informed decisions based on comprehensive data and real-time information.
- **Challenges:** Managing and analyzing large volumes of data requires specialized skills and resources. Data privacy and security concerns also need to be addressed to protect sensitive information.

# The Future of Food Production

## 1. Sustainable Agriculture

### 1.1 Overview

- **Definition:** Sustainable agriculture focuses on practices that balance productivity, environmental health, and social equity. It aims to meet current food needs while preserving resources for future generations.

- **Technologies:** Emerging technologies play a key role in promoting sustainable agriculture by reducing environmental impact, optimizing resource use, and enhancing resilience.

### 1.2 Trends and Innovations

- **Climate-Smart Agriculture:** Technologies such as precision farming and biotechnology support climate-smart agriculture by improving resource efficiency and reducing greenhouse gas emissions.
- **Circular Economy:** The circular economy concept emphasizes reducing waste, recycling resources, and closing nutrient loops. Innovations in waste management and resource recovery contribute to sustainable food production.

### 1.3 Challenges and Opportunities

- **Challenges:** Implementing sustainable practices requires overcoming barriers such as high costs, limited access to technology, and regulatory constraints. Addressing these challenges involves collaboration between stakeholders and supportive policies.
- **Opportunities:** Advances in technology offer opportunities to enhance sustainability through improved resource management, reduced environmental impact, and increased resilience to climate change.

## 2. Food Security and Access

### 2.1 Overview

- **Definition:** Food security refers to the availability, accessibility, and affordability of food for all people. Ensuring food security involves addressing challenges related to production, distribution, and access.
- **Technologies:** Emerging technologies contribute to food security by improving production efficiency, reducing waste, and enhancing supply chain management.

### 2.2 Trends and Innovations

- **Vertical Farming:** Vertical farming and other innovative production methods enable efficient use of space and resources, supporting urban agriculture and local food production.
- **Food Waste Reduction:** Technologies for monitoring and managing food supply chains help reduce food waste and improve distribution efficiency. Solutions include smart packaging, inventory management, and supply chain optimization.

### 2.3 Challenges and Opportunities

- **Challenges:** Addressing food security requires overcoming challenges such as distribution inefficiencies, economic disparities, and supply chain disruptions. Collaborative efforts and innovative solutions are needed to ensure equitable access to food.
- **Opportunities:** Emerging technologies offer opportunities to enhance food security by increasing production efficiency, reducing waste, and improving supply chain management. Innovations in agriculture and food systems contribute to a more resilient and secure food supply.

## 3. Ethical and Social Considerations

### 3.1 Overview

- **Definition:** Ethical and social considerations involve addressing the broader implications of technological advancements in agriculture, including their impact on society, the environment, and animal welfare.
- **Technologies:** Technologies such as biotechnology, robotics, and AI raise ethical questions related to safety, equity, and environmental impact. Ensuring responsible development and implementation of these technologies is crucial.

### 3.2 Trends and Innovations

- **Ethical Frameworks:** Developing ethical frameworks and guidelines for the use of emerging technologies helps address concerns related to safety, equity, and environmental impact.
- **Stakeholder Engagement:** Engaging stakeholders, including farmers, consumers, and policymakers, ensures that technological advancements align with societal values and address public concerns.

### 3.3 Challenges and Opportunities

- **Challenges:** Balancing technological innovation with ethical considerations requires addressing concerns related to safety, equity, and environmental impact. Ensuring transparency and accountability in technology development is essential.
- **Opportunities:** Ethical and social considerations offer opportunities to promote responsible and inclusive technology development. Engaging with diverse stakeholders and adopting ethical practices contribute to sustainable and equitable outcomes.

## Conclusion

Emerging technologies are driving significant changes in agriculture and food production, offering new possibilities for improving productivity, sustainability, and food security. Advances in precision farming, AI, biotechnology, robotics, and digital platforms are transforming agricultural practices and shaping the future of food production. While these technologies offer numerous benefits, they also present challenges that need to be addressed, including cost, accessibility, and ethical considerations. By embracing innovation and addressing these challenges, the agricultural sector can achieve a more resilient, sustainable, and equitable food system. The future of food production lies in harnessing the potential of emerging technologies while ensuring that their development and implementation align with broader societal goals and values.

## References

Borch, K. (2007). Emerging technologies in favour of sustainable agriculture. Futures, 39(9), 1045-1066.

Friha, O., Ferrag, M. A., Shu, L., Maglaras, L., & Wang, X. (2021). Internet of things for the future of smart agriculture: A comprehensive survey of emerging technologies. IEEE/CAA Journal of Automatica Sinica, 8(4), 718-752.

Gokhberg, L., Kuzminov, I., Bakhtin, P., Timofeev, A., & Khabirova, E. (2019). Emerging technologies identification in foresight and strategic planning: case of agriculture and food sector. Emerging Technologies for Economic Development, 205-223.

Newman, L., Newell, R., Dring, C., Glaros, A., Fraser, E., Mendly-Zambo, Z., ... & Kc, K. B. (2023). Agriculture for the Anthropocene: novel applications of technology and the future of food. Food Security, 15(3), 613-627.

# 13

# Regional Approaches to Progressive Agriculture

**Abstract**

*Regional approaches to progressive agriculture involve tailored strategies and practices that address the unique agricultural needs and challenges of specific regions. This chapter explores the various regional strategies for advancing agriculture, focusing on how local conditions, cultures, and resources influence agricultural practices and innovations. It examines case studies from diverse regions to highlight successful approaches and adaptations in progressive agriculture. The chapter covers key areas such as climate adaptation, resource management, technology adoption, and policy frameworks. By analyzing regional successes and challenges, the chapter provides insights into how regional approaches can drive sustainable agricultural development and enhance food security globally.*

***Keywords:*** *Regional agriculture, progressive agriculture, climate adaptation, resource management, technology adoption, policy frameworks, sustainable development, case studies, food security.*

## Introduction

Progressive agriculture refers to innovative and forward-thinking practices aimed at improving agricultural productivity, sustainability, and resilience. While the principles of progressive agriculture are universally applicable, effective implementation often requires region-specific approaches that account for local conditions, resources, and cultural factors. Regional approaches to progressive agriculture involve tailoring strategies to address the unique challenges and opportunities of different areas. By understanding regional strategies and adaptations, stakeholders can develop more targeted and effective solutions for advancing agriculture and ensuring food security.

## Regional Approaches to Progressive Agriculture

### 1. Climate Adaptation

#### Overview

Climate adaptation in agriculture refers to the adjustments and modifications in farming practices and systems to cope with the impacts of climate change.

As global temperatures rise, weather patterns become more unpredictable, and extreme weather events become more frequent, farmers must adapt their practices to maintain productivity and ensure food security. Climate adaptation involves both short-term and long-term strategies that address the specific climatic challenges faced by different regions. Effective adaptation strategies are essential for building resilience in agricultural systems and mitigating the risks associated with climate change.

## Key Strategies for Climate Adaptation

### 1. Crop and Livestock Management

- **Crop Diversification:** Growing a variety of crops reduces dependency on a single crop and spreads risk. Diversification can include the introduction of climate-resilient varieties that are tolerant to heat, drought, or flooding.
- **Livestock Breeding:** Selecting and breeding livestock varieties that are more resilient to changing climatic conditions (e.g., heat stress) can improve productivity and survival rates.
- **Varietal Improvement:** Developing and adopting crop varieties that are resistant to pests, diseases, and extreme weather conditions (e.g., drought-resistant or flood-tolerant varieties) is a key adaptation strategy.

### 2. Water Management

- **Efficient Irrigation:** Implementing efficient irrigation systems such as drip irrigation or sprinkler systems reduces water wastage and ensures that crops receive adequate moisture even during dry periods.
- **Rainwater Harvesting:** Collecting and storing rainwater during wet periods for use during dry spells helps manage water resources more effectively.
- **Soil Moisture Conservation:** Techniques like mulching, conservation tillage, and cover cropping help retain soil moisture and improve soil health.

### 3. Soil Management

- **Conservation Tillage:** Reducing tillage intensity helps maintain soil structure, reduce erosion, and increase organic matter content, which improves soil resilience to extreme weather.
- **Soil Fertility Management:** Applying organic amendments and adopting practices that enhance soil fertility can improve soil health and its ability to withstand climatic stresses.

- **Erosion Control:** Implementing erosion control measures such as contour plowing, terracing, and planting cover crops helps prevent soil loss and degradation.

**4. Agroforestry and Land Use**

- **Agroforestry:** Integrating trees and shrubs into agricultural systems can provide multiple benefits, including windbreaks, improved soil fertility, and enhanced water retention.
- **Land Use Planning:** Adapting land use practices to account for changing climate conditions helps optimize land productivity and reduce vulnerability to climate-related risks.

**5. Technology and Innovation**

- **Climate-Smart Technologies:** Utilizing technologies such as climate forecasting tools, decision support systems, and remote sensing can provide valuable information for managing climatic risks.
- **Genetic Improvement:** Research and development of climate-resilient crops and livestock through biotechnology and genetic engineering enhance adaptability to changing conditions.

**6. Farm Management Practices**

- **Adjusting Planting Dates:** Changing the timing of planting and harvesting to align with shifting climatic patterns can help optimize crop yields and reduce risks associated with weather variability.
- **Pest and Disease Management:** Adapting pest and disease management practices to account for changes in the distribution and prevalence of pests and diseases due to climate change.

## Regional Case Studies

**1. Sub-Saharan Africa**

- **Drought-Resistant Crops:** In Sub-Saharan Africa, the development and adoption of drought-resistant crop varieties, such as improved strains of sorghum and millet, help farmers cope with prolonged dry periods and water scarcity.
- **Community-Based Water Management:** Initiatives that promote community-based water management and rainwater harvesting have improved water availability and resilience in regions experiencing increasing variability in rainfall.

## 2. South Asia

- **Flood-Resilient Varieties:** In South Asia, flood-resilient rice varieties, such as IRRI's "flood-tolerant" rice, have been developed to withstand prolonged submergence and reduce crop losses during flood events.
- **Integrated Crop-Livestock Systems:** Adopting integrated crop-livestock systems helps diversify income sources and improve resilience to climate-induced shocks.

## 3. Mediterranean Region

- **Water-Efficient Irrigation:** In the Mediterranean region, the use of water-efficient irrigation systems such as drip irrigation and the adoption of drought-tolerant crops are key strategies to address water scarcity and maintain agricultural productivity.
- **Soil Conservation Practices:** Practices such as conservation tillage and the use of cover crops help manage soil erosion and improve soil moisture retention.

## Challenges in Climate Adaptation

1. **Financial Constraints:** The cost of implementing adaptation measures can be high, particularly for smallholder and resource-limited farmers. Financial support and investment are necessary to facilitate adaptation efforts.
2. **Technical Knowledge:** Effective adaptation requires technical knowledge and skills that may be lacking in some regions. Training and capacity-building are essential for successful implementation.
3. **Infrastructure Limitations:** Inadequate infrastructure, such as irrigation systems and storage facilities, can hinder the ability to adopt and implement adaptation measures.
4. **Policy and Support:** The effectiveness of adaptation efforts depends on supportive policies and institutional frameworks. Weak policies and lack of support can limit the scope and impact of adaptation initiatives.

## Opportunities for Effective Adaptation

1. **Research and Development:** Investment in research and development of climate-resilient technologies and practices can provide farmers with new tools and solutions for adaptation.
2. **Knowledge Sharing:** Promoting knowledge sharing and collaboration among farmers, researchers, and policymakers can enhance the effectiveness of adaptation strategies and facilitate the dissemination of best practices.

3. **Financial Mechanisms:** Developing financial mechanisms such as insurance schemes, grants, and subsidies can support farmers in implementing adaptation measures and managing climate risks.
4. **Community Engagement:** Engaging local communities in the planning and implementation of adaptation strategies ensures that solutions are tailored to local needs and conditions.

## 2. Resource Management

### 2.1 Overview

- **Definition:** Resource management involves the sustainable use and conservation of natural resources, such as water, soil, and biodiversity. Regional approaches to resource management address local resource constraints and opportunities.
- **Factors Influencing Resource Management:** Regional factors such as land availability, water resources, and soil fertility influence resource management practices.

### 2.2 Case Studies

- **South Asia:** In South Asia, integrated water resource management (IWRM) practices are used to address water scarcity and improve irrigation efficiency. Techniques such as rainwater harvesting, watershed management, and efficient irrigation systems contribute to sustainable resource use.
- **Latin America:** In Latin America, sustainable land management practices, such as agroecology and conservation tillage, are employed to enhance soil fertility and reduce erosion. These practices support long-term agricultural productivity and environmental health.

### 2.3 Challenges and Opportunities

- **Challenges:** Resource management faces challenges such as overexploitation, pollution, and land degradation. Addressing these issues requires effective policies, community engagement, and investment in sustainable practices.
- **Opportunities:** Innovations in resource management, such as precision agriculture and conservation technologies, offer opportunities for improving efficiency and sustainability. Regional collaboration and knowledge-sharing can enhance resource management efforts.

## 3. Technology Adoption

### Overview

Technology adoption in agriculture involves integrating new technological innovations into farming practices to enhance productivity, efficiency, and sustainability. As agricultural challenges evolve—due to factors like climate change, resource constraints, and increasing global food demand—technology plays a crucial role in transforming how farming is conducted. The adoption of advanced technologies can lead to significant improvements in crop yields, resource management, and overall farm management. However, the process of adopting technology varies across regions due to differences in infrastructure, access to resources, and socio-economic conditions.

### Key Technologies

#### 1. Precision Agriculture

- **Definition:** Precision agriculture involves the use of technologies to monitor and manage field variability in crops. This includes technologies like GPS, GIS (Geographic Information Systems), and remote sensing.
- **Applications:** Precision agriculture tools allow for precise application of inputs such as water, fertilizers, and pesticides, based on detailed field data. This leads to optimized resource use, reduced waste, and improved crop health.
- **Impact:** Precision agriculture enhances productivity by tailoring practices to specific field conditions, thus maximizing efficiency and minimizing environmental impact.

#### 2. Artificial Intelligence (AI) and Machine Learning

- **Definition:** AI and machine learning involve using algorithms to analyze large datasets and make predictions or decisions based on patterns and trends.
- **Applications:** In agriculture, AI can be used for predictive analytics, crop health monitoring, and automated decision-making. For instance, AI-powered image recognition systems can identify plant diseases or pests from drone-captured images.
- **Impact:** AI improves decision-making accuracy and enables predictive maintenance, which helps in timely interventions and better management of agricultural operations.

### 3. Biotechnology

- **Definition:** Biotechnology encompasses various techniques to modify and improve crops and livestock through genetic engineering, molecular breeding, and other methods.
- **Applications:** Genetic modification can create crops with desirable traits, such as resistance to pests or tolerance to environmental stresses. CRISPR technology allows for precise gene editing to enhance crop traits or develop disease-resistant plants.
- **Impact:** Biotechnology can increase crop yields, enhance nutritional content, and reduce reliance on chemical inputs, contributing to more sustainable agricultural practices.

### 4. Robotics and Automation

- **Definition:** Robotics and automation involve the use of robots and automated systems to perform agricultural tasks with minimal human intervention.
- **Applications:** Autonomous tractors, robotic harvesters, and drones are examples of agricultural robots. These technologies can perform tasks such as planting, harvesting, and monitoring crops efficiently.
- **Impact:** Robotics and automation reduce labor requirements, increase operational efficiency, and allow for precision farming practices.

### 5. Digital Platforms and Big Data

- **Definition:** Digital platforms and big data involve the use of digital tools and large datasets to collect, analyze, and leverage information for agricultural decision-making.
- **Applications:** Farm management software, data analytics platforms, and cloud-based solutions provide farmers with tools to manage operations, analyze data, and make informed decisions.
- **Impact:** These technologies support smart farming practices by providing insights into crop performance, resource use, and market trends, enabling better management and planning.

## Factors Influencing Technology Adoption

### 1. Infrastructure

- **Access to Technology:** The availability of technological infrastructure, such as internet connectivity, digital devices, and data storage, impacts the adoption of new technologies.

- **Support Systems:** Adequate support systems, including maintenance services and technical support, are essential for successful technology adoption.

## 2. Cost

- **Initial Investment:** High upfront costs for purchasing and implementing new technologies can be a barrier, particularly for smallholder farmers. Cost considerations include the purchase price, installation, and maintenance.
- **Financial Incentives:** Subsidies, grants, and financial support programs can help offset costs and encourage technology adoption.

## 3. Education and Training

- **Skill Development:** Access to education and training is crucial for effective technology adoption. Farmers need to acquire skills to operate and maintain new technologies.
- **Extension Services:** Agricultural extension services and training programs can provide farmers with the knowledge and skills needed to adopt and use new technologies effectively.

## 4. Socio-Economic Conditions

- **Economic Conditions:** Economic stability and access to financial resources influence the ability of farmers to invest in new technologies.
- **Cultural Factors:** Cultural attitudes towards technology and innovation can affect adoption rates. In some regions, traditional practices may be preferred over new technologies.

## 5. Policy and Regulation

- **Supportive Policies:** Government policies and regulations can facilitate or hinder technology adoption. Supportive policies may include incentives for technology adoption, research and development funding, and infrastructure investment.
- **Regulatory Framework:** Clear and transparent regulatory frameworks for new technologies, especially in areas like biotechnology, ensure safe and effective implementation.

## Challenges in Technology Adoption

1. **High Costs:** The initial costs of new technologies can be prohibitive for many farmers, particularly in developing regions.
2. **Technical Complexity:** Some technologies may be complex to use and require specialized skills, which can be a barrier to adoption.

3. **Infrastructure Gaps:** Lack of infrastructure, such as reliable internet access or data storage, can limit the effectiveness of technology implementation.
4. **Resistance to Change:** Cultural resistance and attachment to traditional farming practices can hinder the adoption of new technologies.

### Opportunities for Advancement

1. **Affordable Solutions:** Developing cost-effective and scalable technologies can increase accessibility and adoption rates among smallholder and resource-limited farmers.
2. **Capacity Building:** Investing in education and training programs can empower farmers to adopt and effectively use new technologies.
3. **Public-Private Partnerships:** Collaborations between government, private sector, and research institutions can drive innovation, reduce costs, and support technology adoption.
4. **Innovation and Research:** Continued research and development can lead to new and improved technologies that address specific agricultural challenges and enhance adoption potential.

## 4. Policy Frameworks

### 4.1 Overview

- **Definition:** Policy frameworks involve the development of regulations, incentives, and support mechanisms to guide agricultural practices and development. Regional approaches to policy frameworks address local priorities and challenges.
- **Factors Influencing Policy Frameworks:** Regional policies are influenced by factors such as economic conditions, political contexts, and agricultural priorities.

### 4.2 Case Studies

- **European Union:** The European Union's Common Agricultural Policy (CAP) provides financial support and regulatory frameworks for sustainable farming practices. CAP programs focus on environmental conservation, rural development, and market support.
- **Africa:** In Africa, regional initiatives such as the Comprehensive Africa Agriculture Development Programme (CAADP) aim to enhance agricultural productivity and food security through policy reforms, investment, and regional cooperation.

### 4.3 Challenges and Opportunities

- **Challenges:** Developing effective policy frameworks requires addressing issues such as political instability, limited resources, and conflicting interests. Ensuring stakeholder participation and alignment with regional needs is crucial.
- **Opportunities:** Policy frameworks offer opportunities to support sustainable agricultural development and address regional challenges. Collaborative approaches and evidence-based policy-making can enhance the effectiveness of agricultural policies.

## 5. Socio-Economic Considerations

### 5.1 Overview

- **Definition:** Socio-economic considerations involve understanding the social and economic factors that impact agriculture, including rural livelihoods, labor markets, and access to markets.
- **Factors Influencing Socio-Economic Considerations:** Regional socio-economic conditions, such as income levels, labor availability, and market access, influence agricultural practices and development.

### 5.2 Case Studies

- **Southeast Asia:** In Southeast Asia, initiatives such as rural development programs and microfinance schemes support smallholder farmers by improving access to resources, markets, and credit. These programs enhance livelihoods and promote agricultural development.
- **Eastern Europe:** In Eastern Europe, programs focused on agricultural modernization and rural infrastructure development aim to improve productivity and economic opportunities for farmers. Investments in infrastructure and technology support regional agricultural growth.

### 5.3 Challenges and Opportunities

- **Challenges:** Socio-economic challenges include income disparities, market access barriers, and labor shortages. Addressing these challenges requires targeted interventions and support mechanisms.
- **Opportunities:** Socio-economic programs and investments offer opportunities to enhance livelihoods, improve market access, and support agricultural development. Regional collaboration and inclusive policies can drive positive socio-economic outcomes.

## Conclusion

Regional approaches to progressive agriculture are essential for addressing the diverse challenges and opportunities faced by different areas. By tailoring strategies to local conditions, resources, and socio-economic factors, stakeholders can effectively advance agricultural practices and achieve sustainable development. The case studies presented highlight successful regional strategies and adaptations, offering valuable insights for enhancing agricultural productivity, resource management, and food security. Future efforts should focus on strengthening regional collaboration, investing in technology and infrastructure, and developing policies that support sustainable and inclusive agricultural development. By embracing regional approaches, the agricultural sector can drive innovation, resilience, and growth in diverse contexts, contributing to a more sustainable and food-secure future.

## References

Deshpande, R. S., & Tagade, N. (2021). Progressive Change in Land Reforms and Agriculture Policy. In The Basics of Human Civilization (pp. 157-173). CRC Press.

Ludden, M. T., Welsh, R., Weissman, E., Hilchey, D., Gillespie, G. W., & Guptill, A. (2018). The Progressive Agriculture Index: Assessing the advancement of agri-food systems. Journal of agriculture, food systems, and community development, 8(3), 159-185.

Marsden, T., Munton, R., Ward, N., & Whatmore, S. (1996). Agricultural geography and the political economy approach: a review. Economic geography, 72(4), 361-375.

Wilson, G. A., & Burton, R. J. (2015). 'Neo-productivist'agriculture: Spatio-temporal versus structuralist perspectives. Journal of Rural Studies, 38, 52-64.

# 14

# Integrating Innovations for Sustainable Agriculture

**Abstract**

*Sustainable agriculture aims to meet current food needs while ensuring the health of the environment, economy, and society for future generations. The integration of innovations in technology, practices, and management is crucial for achieving this goal. This chapter explores various innovations that are transforming sustainable agriculture, including advancements in precision farming, biotechnology, renewable energy, and integrated pest management. By analyzing case studies and current research, the chapter demonstrates how these innovations contribute to increased productivity, resource efficiency, and environmental protection. Challenges and opportunities associated with these innovations are also discussed, providing a comprehensive overview of the pathways towards a more sustainable agricultural future.*

***Keywords:*** *Sustainable agriculture, precision farming, biotechnology, renewable energy, integrated pest management, resource efficiency, environmental protection.*

## Introduction

Sustainable agriculture is a dynamic and evolving field that integrates various technological, scientific, and managerial innovations to promote environmental health, economic viability, and social equity. As global challenges such as climate change, resource depletion, and population growth intensify, the need for innovative solutions in agriculture becomes increasingly critical.

## Precision Farming

### Overview

Precision farming, also known as precision agriculture, involves using technology to monitor and manage crop production at a detailed level. This approach aims to optimize inputs and outputs, reduce waste, and enhance productivity. Technologies such as Geographic Information Systems (GIS), Global Positioning Systems (GPS), and remote sensing are central to precision farming.

### Technologies and Techniques

1. **GPS and GIS**: GPS technology enables precise field mapping and tracking, while GIS helps in analyzing spatial data. Together, they allow farmers to implement site-specific management practices, such as variable rate application of fertilizers and pesticides.
2. **Remote Sensing**: This involves the use of satellites or drones to gather data on crop health, soil conditions, and weather patterns. Remote sensing provides real-time information that can be used to make informed decisions about irrigation, nutrient management, and pest control.
3. **Data Analytics**: Advanced data analytics and machine learning algorithms process large datasets from various sources to predict crop yields, optimize resource use, and identify potential issues before they become problems.

### Case Studies

1. **Precision Agriculture in the U.S. Midwest**: Farmers in the U.S. Midwest have adopted precision agriculture techniques to enhance crop yields and reduce input costs. For instance, using GPS-guided tractors and variable rate technology has led to significant improvements in efficiency and productivity.
2. **Precision Rice Farming in Asia**: In countries like China and India, precision farming techniques are being used to manage water use and fertilizer application in rice paddies, leading to increased yields and reduced environmental impact.

### Challenges and Opportunities

While precision farming offers many benefits, it also presents challenges such as high initial costs, the need for technical expertise, and data privacy concerns. However, advancements in technology and decreasing costs are making precision farming more accessible to a broader range of farmers.

## Biotechnology

### Overview

Biotechnology in agriculture involves the use of biological processes and organisms to develop crops with improved traits. This includes genetically modified organisms (GMOs), gene editing, and the development of bio-based products.

### Innovations in Biotechnology

1. **Genetic Engineering**: Genetic modification allows for the insertion of specific genes into crops to confer desirable traits such as resistance to pests, diseases, and environmental stresses.
2. **CRISPR-Cas9**: This gene-editing technology enables precise modifications to the DNA of crops, offering a faster and more targeted approach than traditional genetic engineering.
3. **Biofortification**: This involves breeding crops to increase their nutritional value, addressing micronutrient deficiencies in populations that rely on staple crops.

### Case Studies

1. **Bt Cotton in India**: The introduction of Bt cotton, which is genetically engineered to produce a protein toxic to certain pests, has led to increased yields and reduced pesticide use in India.
2. **Golden Rice**: Genetically modified to produce higher levels of vitamin A, Golden Rice aims to combat vitamin A deficiency in developing countries.

### Challenges and Opportunities

Biotechnology faces challenges including regulatory hurdles, public resistance, and ethical concerns. However, its potential to address food security and nutrition issues is substantial, and ongoing research is likely to overcome many of these barriers.

## Renewable Energy in Agriculture

### Overview

The integration of renewable energy sources in agriculture helps reduce reliance on fossil fuels, lower greenhouse gas emissions, and improve energy efficiency.

### Innovations in Renewable Energy

1. **Solar Power**: Solar panels can be installed on farms to provide electricity for irrigation systems, greenhouse heating, and other farm operations. Solar-powered water pumps and solar greenhouses are examples of practical applications.
2. **Wind Energy**: Wind turbines can generate electricity for farm operations, and their use in rural areas can be particularly beneficial where grid access is limited.

3. **Bioenergy**: This involves using organic materials, such as crop residues and manure, to produce biofuels and biogas. Bioenergy can contribute to reducing waste and providing a renewable source of energy.

## Case Studies

Case studies provide real-world examples of how integrating innovations can enhance sustainable agriculture. They offer insights into successful applications of technologies and practices, highlighting both achievements and lessons learned. Below are detailed case studies from various regions, illustrating the impact of innovations on agricultural sustainability.

### 1. Precision Agriculture in the U.S. Midwest

#### Context

- **Region:** U.S. Midwest, known for its large-scale, industrial agriculture.
- **Innovation:** Precision agriculture technologies, including GPS-guided tractors, yield mapping, and variable rate technology (VRT).

#### Implementation

- **GPS-Guided Tractors:** Farmers use GPS technology to navigate tractors with high accuracy, allowing for precise planting and application of inputs.
- **Yield Mapping:** Yield monitors collect data on crop yields across different field areas, creating detailed yield maps.
- **Variable Rate Technology (VRT):** VRT systems adjust the application rates of fertilizers, pesticides, and water based on real-time field data.

#### Results

- **Increased Yields:** Precision agriculture has led to higher crop yields due to optimized input use and better management practices.
- **Reduced Input Costs:** By applying inputs at variable rates, farmers have reduced overall input costs, such as fertilizers and pesticides.
- **Environmental Impact:** The targeted application of inputs has minimized environmental impacts, such as nutrient runoff and pesticide leaching.
- **Farmer Feedback:** Farmers report improved decision-making, better resource management, and enhanced productivity.

#### Lessons Learned

- **Adoption Barriers:** Initial costs and the need for technical expertise can be barriers to adoption, particularly for small and medium-sized farms.

- **Technology Integration:** Successful integration requires compatible technologies and data management systems.

## 2. Biotechnological Advances in Africa

### Context

- **Region:** Sub-Saharan Africa, with a focus on addressing food security and climate resilience.
- **Innovation:** Development and adoption of drought-resistant genetically modified (GM) maize varieties.

### Implementation

- **GM Maize Varieties:** Researchers developed maize varieties that are resistant to drought, a common issue in arid and semi-arid regions of Africa.
- **Field Trials:** Extensive field trials were conducted to test the performance of these varieties under various climatic conditions.

### Results

- Increased Yields: The adoption of drought-resistant maize has led to higher yields even in regions experiencing severe drought.
- Improved Food Security: Farmers growing these GM varieties have reported increased food production and reduced risk of crop failure.
- Economic Benefits: Higher yields have translated into increased income for farmers, contributing to improved livelihoods.

### Lessons Learned

- **Regulatory Challenges:** The introduction of GM crops faces regulatory and acceptance challenges, which need to be addressed through stakeholder engagement and education.
- **Training and Support:** Providing training and support to farmers is crucial for the successful adoption of biotechnological innovations.

## 3. Renewable Energy in European Farms

The European agricultural sector is increasingly turning to renewable energy sources to enhance sustainability, reduce carbon footprints, and ensure long-term energy security. The integration of renewable energy in European farms is driven by policy incentives, technological advancements, and the need for economic and environmental resilience. This section delves into various renewable energy technologies adopted by European farms, examines case studies of successful implementations, and explores the associated challenges and opportunities.

## Renewable Energy Technologies in European Agriculture

### Solar Power

**Overview**: Solar power is one of the most widely adopted renewable energy technologies in European agriculture. Photovoltaic (PV) systems convert sunlight into electricity, which can be used to power farm operations, irrigation systems, and other equipment.

### Applications:

1. **Solar Panels on Buildings**: Many European farms install solar panels on rooftops of barns, greenhouses, and other structures. This setup not only provides a clean energy source but also helps in reducing electricity bills.
2. **Solar Water Pumps**: Solar-powered water pumps are used for irrigation and livestock watering, particularly in remote areas where traditional power sources are not readily available.

### Case Studies

- **Germany**: In Germany, the "Solar Irrigation Initiative" has seen numerous farms adopting solar-powered irrigation systems. These systems reduce reliance on diesel generators and lower greenhouse gas emissions while maintaining water supply efficiency.
- **Spain**: Spanish vineyards have incorporated solar panels on their buildings and trellises, which helps in managing energy consumption for irrigation and other operations while utilizing the abundant sunlight.

### Wind Energy

**Overview**: Wind energy harnesses the power of wind to generate electricity through wind turbines. While wind turbines are more common in larger-scale operations, smaller turbines are increasingly being used on farms.

### Applications

1. **On-Farm Wind Turbines**: Small to medium-sized wind turbines are installed to generate electricity for farm operations, such as heating, lighting, and machinery.
2. **Combined Wind and Solar Systems**: Some farms combine wind turbines with solar panels to create a hybrid renewable energy system that balances energy supply throughout the year.

### Case Studies

- **Denmark**: Danish farms have been at the forefront of integrating wind energy, with many using small turbines to meet their energy needs. The country's favorable wind conditions make it an ideal location for wind power.
- **The Netherlands**: In the Netherlands, some farms have installed vertical-axis wind turbines, which are less intrusive and can be more suitable for agricultural settings.

### Bio-energy

**Overview**: Bio-energy involves the production of energy from organic materials, including crop residues, manure, and other agricultural by-products. This can be converted into biofuels, biogas, or other energy forms.

### Applications

1. **Biogas Production**: Anaerobic digesters convert manure and organic waste into biogas, which can be used for heating, electricity generation, and as a fuel for vehicles.
2. **Biofuel Production**: Crop residues and dedicated energy crops are processed into biofuels like biodiesel or ethanol, which can replace fossil fuels in farm machinery.

### Case Studies

- **Italy**: Italian farms utilize biogas plants to manage manure and crop residues. The biogas produced is used for electricity generation and heating, contributing to energy independence and waste reduction.
- **France**: In France, some farms have established small-scale biofuel production units that convert used cooking oil and agricultural residues into biodiesel for farm machinery.

### Benefits of Renewable Energy in European Farms

1. **Reduction in Carbon Emissions**: By switching to renewable energy sources, European farms significantly cut their greenhouse gas emissions, contributing to climate change mitigation.
2. **Energy Independence**: Renewable energy reduces reliance on external energy sources, providing farms with greater energy security and stability.
3. **Cost Savings**: Over time, renewable energy systems can lead to lower energy costs. While initial investments can be high, savings on energy bills and government incentives often offset these costs.

4. **Environmental Stewardship**: Adopting renewable energy supports sustainable farming practices, enhancing the overall environmental quality of farm operations.

## Challenges and Opportunities

### Challenges

1. **High Initial Costs**: The upfront investment for renewable energy systems can be substantial. Although long-term savings are significant, the initial capital required can be a barrier for some farms.
2. **Technological Integration**: Integrating renewable energy technologies with existing farm infrastructure can be complex, requiring careful planning and sometimes specialized expertise.
3. **Intermittency of Energy Supply**: Renewable energy sources like solar and wind are subject to variability. Farms need to manage energy storage and backup systems to ensure a reliable supply.

### Opportunities

1. **Policy Incentives**: European Union policies and national regulations often provide subsidies, tax credits, and grants for renewable energy projects, making it financially more attractive for farms to invest in these technologies.
2. **Technological Advancements**: Ongoing research and development in renewable energy technologies are improving efficiency and reducing costs, making them more accessible for farms.
3. **Climate Goals**: The European Union's commitment to achieving climate goals and sustainability targets creates a favorable environment for renewable energy adoption in agriculture.

## Future Trends

1. **Hybrid Systems**: The combination of multiple renewable energy sources, such as solar-wind hybrids or solar-bioenergy systems, is likely to become more prevalent, providing more reliable and efficient energy solutions.
2. **Energy Storage**: Advances in energy storage technologies, such as batteries and thermal storage systems, will enhance the reliability of renewable energy supply on farms.
3. **Smart Grid Integration**: The development of smart grids and energy management systems will enable better integration of renewable energy sources into farm operations and the broader energy grid.

## 4. Soil Health Management in Latin America

### Context

- **Region:** Latin America, where soil degradation and erosion are major challenges.
- **Innovation:** Implementation of conservation tillage and cover cropping practices.

### Implementation

- **Conservation Tillage:** Farmers adopted reduced tillage methods to maintain soil structure and organic matter.
- **Cover Cropping:** Cover crops were planted between main crop cycles to improve soil fertility and reduce erosion.

### Results

- **Enhanced Soil Fertility:** Soil health improved with increased organic matter and better nutrient availability.
- **Reduced Erosion:** Conservation tillage and cover cropping have led to reduced soil erosion and improved water retention.
- **Increased Productivity:** Healthier soils have contributed to higher crop yields and improved farm productivity.

### Lessons Learned

- **Long-Term Commitment:** Soil health management practices require a long-term commitment and continuous adaptation.
- **Knowledge Transfer:** Effective implementation depends on the transfer of knowledge and best practices to farmers.

## 5. Circular Economy Practices in Asia

### Context

- **Region:** Various Asian countries, with a focus on integrating circular economy principles into agriculture.
- **Innovation:** Adoption of waste recycling and resource efficiency practices.

### Implementation

- **Waste Recycling:** Agricultural waste, such as crop residues and livestock manure, is recycled into compost and bioenergy.
- **Resource Efficiency:** Farmers implemented practices to minimize waste and make the most of available resources.

### Results

- **Reduced Waste:** Recycling agricultural waste has minimized waste and contributed to resource conservation.
- **Improved Soil Health:** Composting and the use of recycled organic matter have improved soil fertility and health.
- **Sustainable Resource Management:** Circular economy practices have enhanced resource efficiency and supported sustainable agricultural systems.

### Lessons Learned

- **Infrastructure Needs:** Successful implementation requires appropriate infrastructure for waste collection, processing, and recycling.
- **Farmer Participation:** Engaging farmers in circular economy practices requires education and incentives to participate.

### Challenges and Opportunities

While renewable energy technologies offer significant benefits, they also face challenges such as high initial investment costs and technological limitations. However, ongoing advancements and supportive policies are helping to make these technologies more viable and widespread.

## Integrated Pest Management (IPM)

### Overview

Integrated Pest Management (IPM) is a holistic approach to managing pests that combines multiple strategies to minimize their impact on crops while reducing reliance on chemical pesticides.

### Innovations in IPM

1. **Biological Control**: The use of natural predators or parasites to control pest populations can reduce the need for chemical pesticides and promote ecological balance.
2. **Cultural Practices**: Crop rotation, planting pest-resistant varieties, and optimizing planting schedules are cultural practices that help manage pest populations.
3. **Pheromone Traps**: These traps use pheromones to attract and capture pests, providing a method of monitoring and controlling pest populations without harmful chemicals.

## Case Studies

1. **Cotton IPM in the U.S.**: The adoption of IPM strategies in cotton farming has led to reduced pesticide use and increased sustainability.
2. **Rice IPM in Asia**: Integrated pest management practices in rice cultivation have improved pest control and reduced the environmental impact of rice farming.

## Challenges and Opportunities

IPM faces challenges such as the need for farmer education and the complexity of managing multiple pest control methods. However, the benefits of reduced chemical use and improved environmental health make IPM a promising approach for sustainable agriculture.

## Conclusion

Integrating innovations in precision farming, biotechnology, renewable energy, and integrated pest management represents a multifaceted approach to achieving sustainable agriculture. Each of these innovations contributes to improving productivity, resource efficiency, and environmental stewardship. While challenges remain, the continued advancement and adoption of these technologies hold great promise for creating a more sustainable and resilient agricultural system

## References

.Sulochna, M. Z., Patel, A. K., Kumar, N., & Venkateswarlu, M. (2023). Innovations in Sustainable Agriculture: Integrating Technology and Traditional Practices for Crop Improvement. Journal of Plant Biota.

Farooq, M., & Pisante, M. (Eds.). (2019). Innovations in sustainable agriculture (pp. 1-627). Berlin/Heidelberg, Germany: Springer International Publishing.

Loconto, A. M., Poisot, A. S., & Santacoloma, P. (2016). Innovative markets for sustainable agriculture.

Pant, L. P., & Hambly-Odame, H. (2009). Innovations systems in renewable natural resource management and sustainable agriculture: a literature review. African Journal of Science, Technology, Innovation and Development, 1(1), 103-135.

# 15

# The Role of Stakeholders in Advancing Agriculture

**Abstract**

*Agriculture is a critical sector that impacts global food security, economic stability, and environmental sustainability. The advancement of agriculture relies heavily on the active involvement and collaboration of various stakeholders, including farmers, government agencies, private sector players, non-governmental organizations (NGOs), and academic institutions. This chapter explores the multifaceted roles of these stakeholders in driving agricultural innovation, policy development, and sustainable practices. Through an analysis of stakeholder interactions and contributions, the chapter highlights key areas where collaboration is essential for advancing agriculture. Case studies and real-world examples are used to illustrate the impact of stakeholder engagement on agricultural outcomes. The chapter concludes by discussing the challenges and opportunities associated with stakeholder collaboration and offers recommendations for enhancing collective efforts in agriculture.*

***Keywords:*** *Agriculture, stakeholders, innovation, policy development, sustainable practices, collaboration, food security, economic stability, environmental sustainability.*

## Introduction

Agriculture is a vital sector that influences global food security, economic growth, and environmental sustainability. The advancement of agriculture requires the collective efforts of various stakeholders, each contributing unique resources, expertise, and perspectives. Understanding the roles and interactions of these stakeholders is crucial for developing effective strategies to address agricultural challenges and seize opportunities for growth. This chapter examines the roles of different stakeholders in advancing agriculture, explores their contributions and collaborations, and discusses the challenges and opportunities associated with their involvement.

## Stakeholders in Agriculture

### 1. Farmers

#### Role and Contributions

- **Innovators**: Farmers are at the forefront of adopting new technologies and practices. Their direct involvement in agriculture allows them to experiment with and implement innovations such as precision farming, organic practices, and climate-smart agriculture.
- **Feedback Providers**: Farmers provide valuable feedback on the effectiveness and practicality of new technologies and policies, helping to refine and improve agricultural solutions.

#### Challenges

- **Access to Resources**: Farmers may face challenges in accessing financial resources, technology, and information necessary for innovation.
- **Risk Management**: The volatility of agricultural markets and climate change can create significant risks for farmers, impacting their ability to adopt new practices.

### 2. Government Agencies

#### Role and Contributions

- **Policy Development**: Government agencies are responsible for creating and enforcing policies that shape agricultural practices. This includes regulations on food safety, environmental protection, and subsidies.
- **Funding and Support**: Governments provide funding for research, infrastructure development, and extension services that support agricultural innovation and sustainability.

#### Challenges

- **Policy Implementation**: Effective implementation of agricultural policies can be hampered by bureaucratic hurdles and lack of coordination among agencies.
- **Balancing Interests**: Governments must balance the needs of various stakeholders, including farmers, consumers, and environmental groups, which can be challenging.

### 3. Private Sector

#### Role and Contributions

- **Technology and Innovation**: Private companies invest in research and development to create new technologies, such as genetically modified crops, advanced irrigation systems, and digital tools for farm management.

- **Market Access**: Private sector players, including agribusinesses and food processors, provide farmers with access to markets and value-added services.

**Challenges**

- **Profit Motive**: Private sector interests may not always align with public goals, such as environmental sustainability or social equity.
- **Market Concentration**: The concentration of market power among a few large companies can limit competition and reduce opportunities for smaller producers.

**4. Non-Governmental Organizations (NGOs)**

**Role and Contributions**

- **Advocacy and Awareness**: NGOs play a crucial role in advocating for sustainable agricultural practices, environmental protection, and social equity. They raise awareness about issues such as land rights, climate change, and food security.
- **Capacity Building**: NGOs provide training, technical assistance, and resources to farmers and communities, helping them adopt sustainable practices and improve their livelihoods.

**Challenges**

- **Funding Constraints**: NGOs often rely on donations and grants, which can be unpredictable and limit their capacity to implement large-scale projects.
- **Coordination**: Effective coordination with other stakeholders, including governments and private sector players, is essential but can be challenging.

**5. Academic Institutions**

**Role and Contributions**

- **Research and Development**: Academic institutions conduct research on agricultural practices, crop varieties, pest management, and sustainability. Their work provides the scientific basis for innovations and policy recommendations.
- **Education and Training**: Universities and research institutes offer education and training programs for students, farmers, and extension agents, building the knowledge and skills needed for modern agriculture.

## Challenges

- **Application of Research**: Bridging the gap between academic research and practical application in the field can be difficult.
- **Funding and Resources**: Research institutions may face challenges in securing adequate funding and resources for long-term projects.

## Collaborative Efforts and Case Studies

### 1. Public-Private Partnerships (PPPs)

**Overview**: Public-private partnerships (PPPs) involve collaboration between government agencies and private sector companies to address agricultural challenges and promote innovation. These partnerships can leverage the strengths of both sectors to achieve common goals.

#### Case Study

- **The CGIAR System**: The Consultative Group on International Agricultural Research (CGIAR) is a global partnership involving governments, research institutions, and private organizations. CGIAR focuses on agricultural research and development, with initiatives aimed at improving crop productivity, resilience, and sustainability. Its collaborative approach has led to significant advances in crop varieties and farming practices.

### 2. Farmer Cooperatives

**Overview**: Farmer cooperatives are organizations formed by groups of farmers who work together to achieve common goals, such as improving access to markets, sharing resources, and enhancing bargaining power.

#### Case Study

- **The Organic Valley Cooperative**: Based in the United States, Organic Valley is a cooperative of organic dairy farmers. By pooling resources and leveraging collective marketing efforts, the cooperative has successfully promoted organic dairy products and supported sustainable farming practices.

### 3. NGO-Government Partnerships

**Overview**: Partnerships between NGOs and government agencies can address specific agricultural challenges, such as improving food security, promoting sustainable practices, and supporting rural development.

## Case Study

- **The FAO's Work with NGOs**: The Food and Agriculture Organization (FAO) of the United Nations collaborates with various NGOs to implement projects focused on food security, nutrition, and sustainable agriculture. These partnerships have led to successful initiatives in areas such as smallholder farmer support and disaster resilience.

## Challenges in Stakeholder Collaboration

### 1. Misaligned Objectives

Different stakeholders often have varying objectives and priorities, which can lead to conflicts and challenges in achieving common goals. For example, private sector companies may prioritize profitability, while NGOs may focus on environmental and social outcomes.

### 2. Communication and Coordination

Effective communication and coordination among stakeholders are essential for successful collaboration. However, differing interests, organizational cultures, and communication barriers can hinder these efforts.

### 3. Resource Constraints

Limited resources, including funding, expertise, and infrastructure, can impact the ability of stakeholders to collaborate effectively and implement projects.

## Opportunities for Enhancing Stakeholder Collaboration

Enhancing stakeholder collaboration is crucial for advancing agricultural innovation, sustainability, and resilience. Effective collaboration among farmers, government agencies, private sector entities, non-governmental organizations (NGOs), and academic institutions can address complex agricultural challenges and leverage collective strengths. This section explores various opportunities for improving stakeholder collaboration in agriculture, focusing on building trust and relationships, leveraging technology, and strengthening policy frameworks.

### 1. Building Trust and Relationships

### 1.1. Establishing Common Goals

**Opportunity**: Identifying and articulating common goals is foundational for effective collaboration. By aligning objectives, stakeholders can work together more cohesively and achieve mutually beneficial outcomes.

**Examples**

- **Sustainable Farming Initiatives**: Joint efforts to promote sustainable farming practices, such as soil conservation or water management, can bring together diverse stakeholders with shared environmental and economic goals.
- **Food Security Projects**: Collaborative projects aimed at enhancing food security, such as improving supply chains or developing resilient crop varieties, benefit from a unified vision among stakeholders.

### 1.2 Fostering Open Communication

**Opportunity**: Open and transparent communication helps build trust and facilitates better understanding among stakeholders. Regular updates, shared information, and active listening are essential components.

**Examples**

- **Regular Meetings and Workshops**: Holding regular meetings and workshops allows stakeholders to discuss progress, share insights, and address concerns.
- **Collaborative Platforms**: Using online platforms for communication and data sharing can streamline interactions and ensure that all stakeholders are informed and engaged.

### 1.3 Building Long-Term Relationships

**Opportunity**: Long-term relationships based on mutual respect and shared interests contribute to more sustainable and effective collaboration. Investing in relationship-building activities can yield long-term benefits.

**Examples**

- **Partnership Agreements**: Formal agreements or memorandums of understanding (MOUs) can define roles, responsibilities, and expectations, helping to build lasting partnerships.
- **Joint Ventures and Initiatives**: Engaging in joint ventures or collaborative initiatives creates opportunities for stakeholders to work together on specific projects or programs.

## 2. Leveraging Technology

### 2.1 Utilizing Digital Tools

**Opportunity**: Digital tools and technologies can enhance collaboration by providing platforms for communication, data sharing, and project management. These tools can improve efficiency and coordination among stakeholders.

**Examples**

- **Project Management Software**: Tools like Asana or Trello help manage collaborative projects by tracking tasks, deadlines, and responsibilities.
- **Data Analytics Platforms**: Platforms that aggregate and analyze agricultural data can provide insights for decision-making and foster collaboration by sharing valuable information.

### 2.2 Implementing Geographic Information Systems (GIS)

**Opportunity**: GIS technology allows stakeholders to visualize and analyze spatial data related to agriculture, such as land use, crop health, and resource distribution. This shared understanding can improve coordination and planning.

**Examples**

- **Precision Agriculture**: GIS tools used in precision agriculture enable farmers, researchers, and policymakers to collaborate on optimizing resource use and improving crop management.
- **Disaster Response**: GIS can be used to coordinate responses to agricultural disasters, such as floods or droughts, by providing real-time information on affected areas and resource needs.

### 2.3 Encouraging Innovation through Technology Hubs

**Opportunity**: Technology hubs and innovation centers can bring together stakeholders to collaborate on developing and implementing new agricultural technologies and practices.

**Examples**

- **Agri-Tech Incubators**: Agri-tech incubators support startups and entrepreneurs in developing innovative solutions for agriculture. Collaboration among these startups, investors, and research institutions can drive technological advancements.
- **Innovation Labs**: Innovation labs that focus on agricultural challenges provide a space for stakeholders to test and refine new technologies and practices.

## 3. Strengthening Policy Frameworks

### 3.1. Developing Supportive Policies

**Opportunity**: Creating and implementing policies that encourage stakeholder collaboration can enhance collective efforts and drive positive outcomes in agriculture.

### Examples

- **Incentives for Collaboration**: Policies that provide financial or regulatory incentives for collaborative projects can motivate stakeholders to work together on agricultural innovations and sustainability initiatives.
- **Funding Programs**: Government funding programs that support collaborative research and development projects can help stakeholders pool resources and expertise.

## 3.2. Facilitating Public-Private Partnerships (PPPs)

**Opportunity**: Public-private partnerships (PPPs) can leverage the strengths of both sectors to address agricultural challenges and promote innovation. PPPs can provide resources, expertise, and infrastructure for collaborative projects.

### Examples

- **Research Collaborations**: PPPs involving government agencies, private companies, and research institutions can drive research and development in areas such as crop improvement, pest management, and sustainable practices.
- **Infrastructure Development**: Collaborative efforts to develop infrastructure, such as irrigation systems or processing facilities, can benefit from the combined resources and expertise of public and private stakeholders.

## 3.3. Enhancing Policy Coordination

**Opportunity**: Coordinating policies across different levels of government and sectors can improve alignment and effectiveness in addressing agricultural challenges.

### Examples

- **Integrated Policy Frameworks**: Developing integrated policy frameworks that address agricultural, environmental, and economic goals can enhance coordination among stakeholders and promote comprehensive solutions.
- **Cross-Sectoral Collaboration**: Encouraging collaboration between agricultural, environmental, and economic policymakers can lead to more effective and holistic approaches to agricultural challenges.

## 4. Engaging in Collaborative Research and Development

### 4.1. Joint Research Initiatives

**Opportunity**: Collaborative research initiatives involving multiple stakeholders can accelerate the development and adoption of innovative agricultural technologies and practices.

**Examples**

- **Multi-Stakeholder Research Projects**: Research projects that involve farmers, researchers, and industry experts can address specific agricultural challenges and generate practical solutions.
- **Field Trials and Demonstrations**: Collaborative field trials and demonstrations allow stakeholders to test and evaluate new technologies and practices in real-world settings.

### 4.2 Knowledge Sharing and Capacity Building

**Opportunity**: Sharing knowledge and building capacity among stakeholders can enhance their ability to collaborate effectively and implement innovative solutions.

**Examples**

- **Training Programs**: Joint training programs and workshops can provide stakeholders with the skills and knowledge needed to adopt new technologies and practices.
- **Knowledge Networks**: Creating networks for knowledge sharing, such as forums or communities of practice, can facilitate collaboration and information exchange among stakeholders.

## 5. Promoting Inclusivity and Equity

### 5.1. Engaging Diverse Stakeholders

**Opportunity**: Ensuring that diverse stakeholders, including marginalized and underrepresented groups, are engaged in agricultural collaboration can lead to more inclusive and equitable outcomes.

**Examples**

- **Community Involvement**: Involving local communities and smallholder farmers in decision-making processes can ensure that their needs and perspectives are considered.
- **Gender Equity**: Promoting gender equity in agricultural collaboration can enhance the participation and contributions of women and other underrepresented groups.

## 5.2 Addressing Equity Issues

**Opportunity**: Addressing equity issues and ensuring that the benefits of collaboration are distributed fairly can enhance the effectiveness and sustainability of agricultural initiatives.

### Examples

- **Resource Allocation**: Ensuring equitable allocation of resources, such as funding and technology, can help address disparities and support disadvantaged stakeholders.
- **Policy Support**: Developing policies that promote equity and inclusion in agricultural collaboration can contribute to more balanced and effective outcomes.

## Conclusion

The advancement of agriculture relies on the active involvement and collaboration of various stakeholders, including farmers, government agencies, private sector players, NGOs, and academic institutions. Each stakeholder plays a unique role in driving innovation, policy development, and sustainable practices. By understanding and addressing the challenges and opportunities associated with stakeholder collaboration, it is possible to achieve more effective and impactful outcomes in agriculture. Building strong relationships, leveraging technology, and strengthening policy frameworks are essential for enhancing collective efforts and advancing the agricultural sector towards a more sustainable and resilient future.

## References

Bhattacharyya, T., Wani, S. P., & Tiwary, P. (2021). Empowerment of stakeholders for scaling-up: digital technologies for agricultural extension. Scaling-up solutions for farmers: technology, partnerships and convergence, 121-147.

Hellin, J., Balié, J., Fisher, E., Blundo-Canto, G., Meah, N., Kohli, A., & Connor, M. (2020). Sustainable agriculture for health and prosperity: stakeholders' roles, legitimacy and modus operandi. Development in Practice, 30(7), 965-971.

Reimer, A., Doll, J. E., Boring, T. J., & Zimnicki, T. (2023). Scaling up conservation agriculture: An exploration of challenges and opportunities through a stakeholder engagement process. Journal of Environmental Quality, 52(3), 465-475.

Rose, D. C., Parker, C., Fodery, J., Park, C., Sutherland, W. J., & Dicks, L. V. (2018). Involving stakeholders in agricultural decision support systems: Improving user-centred design. International Journal of Agricultural Management, 6, 80-89.

# 16

# Vision for the Future of Progressive Farming

**Abstract**

*The future of progressive farming is shaped by the convergence of technology, sustainability, and innovation. This chapter explores the emerging trends, technologies, and strategies that are expected to define the next generation of farming practices. By integrating advancements in precision agriculture, biotechnology, sustainable resource management, and digital transformation, progressive farming aims to enhance productivity, sustainability, and resilience in agriculture. This vision encompasses the use of cutting-edge technologies, the adoption of holistic and sustainable practices, and the role of policy and education in fostering a new era of agriculture. Case studies and projections highlight the potential pathways and challenges for achieving these future goals.*

***Keywords***: *Progressive farming, future agriculture, precision agriculture, biotechnology, sustainability, digital transformation, agricultural technology, resilience, policy, education.*

## Introduction

The landscape of agriculture is undergoing a transformative shift, driven by the need to address global challenges such as climate change, resource scarcity, and a growing population. The future of farming is increasingly characterized by progressive practices that leverage technology, sustainability, and innovation to enhance productivity, resilience, and environmental stewardship. This vision for the future of progressive farming integrates a range of advanced technologies, sustainable practices, and holistic strategies to create a more efficient and sustainable agricultural system.

## The Need for Change

### Global Challenges

- **Population Growth:** The global population is projected to reach nearly 10 billion by 2050, increasing the demand for food and necessitating significant enhancements in agricultural productivity and efficiency.

- **Climate Change:** Climate change is impacting agricultural systems through altered weather patterns, increased frequency of extreme weather events, and shifting growing seasons. Farmers need to adapt to these changes to maintain food security and productivity.
- **Resource Scarcity:** The availability of natural resources such as water, soil, and arable land is diminishing. Efficient resource management and sustainable practices are essential to ensure the long-term viability of agriculture.

### Technological Advancements

- **Precision Agriculture:** Technologies such as GPS, remote sensing, and data analytics are revolutionizing farm management by enabling more precise monitoring and management of crops and resources. Precision agriculture allows for optimized input use, reduced waste, and improved crop yields.
- **Biotechnology:** Innovations in genetic engineering and molecular breeding are creating crops with enhanced traits such as resistance to pests and diseases, improved nutritional content, and better resilience to environmental stresses.
- **Digital Tools:** The integration of digital technologies, including the Internet of Things (IoT), blockchain, and big data analytics, is transforming agricultural operations by providing real-time data, enhancing supply chain transparency, and improving decision-making.

## The Vision for Progressive Farming

### Sustainability and Efficiency

- **Resource Management:** Progressive farming emphasizes the efficient use of natural resources, including water, soil, and energy. Sustainable practices such as conservation tillage, cover cropping, and integrated pest management are integral to maintaining soil health, reducing environmental impact, and optimizing resource use.
- **Renewable Energy:** The adoption of renewable energy sources, such as solar power, wind energy, and biogas, is reducing the carbon footprint of farming operations and enhancing energy independence. Renewable energy technologies are being integrated into farm infrastructure to power operations and manage waste.

### Innovation and Integration

- **Technology Integration:** The future of progressive farming involves the seamless integration of various technologies to create a cohesive and efficient agricultural system. This includes combining precision agriculture tools with biotechnology innovations and digital platforms to enhance farm productivity and sustainability.

- **Holistic Management:** Progressive farming adopts a holistic approach that considers the interactions between various components of the agricultural system. This includes integrating crop and livestock management, optimizing input use, and addressing environmental and social impacts.

## Policy and Education

- **Supportive Policies:** Effective policies and regulations are crucial for supporting the adoption of innovative technologies and sustainable practices. Governments and organizations can provide financial incentives, grants, and subsidies to facilitate the transition to progressive farming.
- **Education and Training:** Ensuring that farmers and stakeholders have access to education and training is essential for successful implementation. Knowledge transfer and capacity building will empower farmers to adopt and effectively use new technologies and practices.

## The Path Forward

## Challenges and Opportunities

- **Financial Constraints:** The initial investment required for adopting new technologies and practices can be a barrier, particularly for small and medium-sized farms. Addressing financial constraints through incentives and subsidies is crucial for widespread adoption.
- **Technical Expertise:** The need for specialized skills and knowledge to manage and integrate advanced technologies is a significant challenge. Providing training and support will enable farmers to effectively use and benefit from these innovations.

## Collaboration and Innovation

- **Research and Development:** Continued investment in research and development will drive technological advancements and create new opportunities for sustainable farming. Collaboration between researchers, policymakers, and farmers is essential for advancing agricultural innovation.
- **Global and Local Integration:** The future of progressive farming will involve integrating global innovations with local practices and conditions. Tailoring solutions to specific regional challenges and opportunities will enhance the effectiveness of agricultural practices.

## Emerging Trends and Technologies

### 1. Precision Agriculture

#### 1.1 Overview

- **Definition:** Precision agriculture involves using technology to monitor and manage field variability in crops, optimizing resource use and improving yields.
- **Technologies:** Key technologies include GPS, GIS, remote sensing, and data analytics.

#### 1.2 Future Directions

- **Enhanced Sensors and Data Collection:** Advances in sensor technology will enable more precise and detailed data collection, improving decision-making and resource management.
- **AI and Machine Learning:** Artificial intelligence and machine learning algorithms will enhance predictive analytics, enabling farmers to anticipate and respond to crop and environmental conditions more effectively.

#### 1.3 Impact

- **Increased Efficiency:** Precision agriculture will continue to optimize input use, reduce waste, and improve crop yields.
- **Environmental Benefits:** By minimizing resource use and environmental impact, precision agriculture will contribute to more sustainable farming practices.

### 2. Biotechnology

#### 2.1 Overview

- **Definition:** Biotechnology involves using biological processes and organisms to develop or enhance agricultural products and practices.
- **Techniques:** Key techniques include genetic modification, molecular breeding, and CRISPR gene editing.

#### 2.2 Future Directions

- **Advanced Genetic Engineering:** Innovations in gene editing will enable the development of crops with enhanced traits such as drought tolerance, disease resistance, and improved nutritional content.
- **Synthetic Biology:** Synthetic biology will allow for the creation of novel crops and microorganisms designed to address specific agricultural challenges.

### 2.3 Impact

- **Enhanced Crop Resilience:** Biotechnology will contribute to the development of crops that can withstand environmental stresses and improve food security.
- **Nutritional Improvement:** Advances in biotechnology will enhance the nutritional value of crops, addressing deficiencies and improving public health.

## 3. Digital Transformation

### 3.1 Overview

- **Definition:** Digital transformation involves the integration of digital technologies into all aspects of agricultural operations.
- **Technologies:** Key technologies include IoT, blockchain, big data analytics, and cloud computing.

### 3.2 Future Directions

- **IoT and Smart Farming:** The Internet of Things (IoT) will enable real-time monitoring and automation of farming operations, enhancing efficiency and productivity.
- **Blockchain for Transparency:** Blockchain technology will provide greater transparency and traceability in the supply chain, improving food safety and quality.

### 3.3 Impact

- **Operational Efficiency:** Digital tools will streamline farm operations, reduce labor costs, and enhance decision-making.
- **Supply Chain Management:** Digital transformation will improve the efficiency and transparency of the agricultural supply chain.

## 4. Sustainable Resource Management

### 4.1 Overview

- **Definition:** Sustainable resource management involves practices that conserve and optimize the use of natural resources while minimizing environmental impact.
- **Practices:** Key practices include water conservation, soil health management, and sustainable energy use.

### 4.2 Future Directions

- **Water Efficiency:** Advanced irrigation technologies and water management practices will enhance water use efficiency and reduce wastage.
- **Soil Health Innovations:** New practices and technologies will improve soil health, fertility, and carbon sequestration.

### 4.3 Impact

- **Resource Conservation:** Sustainable resource management will ensure the long-term availability of essential resources and reduce environmental impact.
- **Enhanced Resilience:** Improved resource management practices will increase the resilience of farming systems to environmental changes and stresses.

## Case Studies

### 1. Smart Farming in the Netherlands

- **Context:** The Netherlands is a leader in smart farming technologies, with a focus on precision agriculture and digital transformation.
- **Implementation:** Dutch farms utilize advanced sensors, automated systems, and data analytics to optimize crop production and resource use.
- **Impact:** Increased productivity, reduced environmental impact, and enhanced operational efficiency.

### 2. CRISPR Technology in U.S. Agriculture

- **Context:** CRISPR gene editing technology is being used to develop crops with enhanced traits.
- **Implementation:** U.S. researchers are using CRISPR to create crops with improved drought resistance and disease resistance.
- **Impact:** Enhanced crop resilience, increased yields, and improved food security.

### 3. Blockchain in Supply Chain Management

- **Context:** Blockchain technology is being adopted to improve transparency and traceability in the agricultural supply chain.
- **Implementation:** Blockchain systems are used to track the origin and movement of agricultural products from farm to table.

- **Impact:** Improved food safety, reduced fraud, and enhanced supply chain efficiency.

## Policy and Regulation

### 1. Supporting Innovation

- **Incentives:** Governments can provide financial incentives, grants, and subsidies to support the adoption of innovative technologies and sustainable practices.
- **Regulatory Framework:** Developing a supportive regulatory framework is essential for facilitating the implementation of new technologies and practices.

### 2. Promoting Sustainability

- **Sustainability Standards:** Establishing and enforcing sustainability standards can ensure that agricultural practices meet environmental and social criteria.
- **Education and Training:** Providing education and training programs for farmers is crucial for successful adoption and implementation of new technologies.

## Education and Knowledge Transfer

### 1. Capacity Building

- **Training Programs:** Developing training programs and resources to enhance farmers' skills and knowledge in using advanced technologies and sustainable practices.
- **Knowledge Sharing:** Facilitating knowledge sharing and collaboration between researchers, policymakers, and farmers to promote best practices and innovation.

### 2. Future Skills

- **Technical Expertise:** Ensuring that the next generation of farmers is equipped with the technical expertise needed to manage and integrate advanced technologies.
- **Adaptability:** Fostering adaptability and innovation skills to navigate the evolving landscape of agriculture.

## Challenges and Opportunities

### 1. Challenges

- **Financial Barriers:** High costs of new technologies and practices can be a barrier to adoption, particularly for small and medium-sized farms.

- **Technical Expertise:** The need for specialized skills and knowledge to effectively implement and manage new technologies.

**2. Opportunities**

- **Innovation:** Continued investment in research and development can drive technological advancements and create new opportunities for sustainable farming.
- **Collaboration:** Public-private partnerships and collaborative initiatives can enhance technology transfer and support the adoption of innovative practices.

## Conclusion

The vision for the future of progressive farming involves a comprehensive integration of technology, sustainability, and innovation. By embracing advancements in precision agriculture, biotechnology, digital transformation, and sustainable resource management, the agricultural sector can achieve greater productivity, resilience, and environmental stewardship. Successful implementation will require addressing challenges such as financial constraints and technical expertise, while capitalizing on opportunities for research, collaboration, and policy support. As the agricultural landscape continues to evolve, the future of progressive farming holds the promise of a more sustainable and efficient agricultural system that meets the needs of both current and future generations

## References

Duke, R.K. (2020). From Dewey to Skyline Farms: Progressive Education in New Deal Resettlement Communities (Doctoral dissertation, Middle Tennessee State University).

Harper, D. (2001). Changing works: Visions of a lost agriculture. University of Chicago Press.

Klinkenbergh, O.A. (2021). Framing Agroecology in the Netherlands: Identifying strategies being used by progressive farmers and farming organisations and how policy and scientific debates align with them (Master›s thesis).

Thomas, R.Z. (2011). Tilling new soil: coverage of organic agriculture in farm journal, successful farming, and progressive farmer from 1985 to 2005 (Master›s thesis, Ohio University).